MW01342211

"Katie brings a fresh, thoughtful energy to cocktails. This book feels just like her: creative, approachable, and full of great ideas."

—**Natalie Migliarini**, author of *Beautiful Booze: Stylish Cocktails to Make at Home*

"Stryjewski has created a truly delightful combination of ornithology and cocktails. Her storytelling brings bird-related natural history to life with each familiar yet creative recipe. As an ornithologist, I'm impressed with the level of detail she has managed to convey in such an accessible way. Everyone from experienced birdwatchers to cocktail enthusiasts is sure to find something to learn and enjoy."

—**Allison Shultz**, Associate Curator of Ornithology at Natural History Museum of Los Angeles County

"What a fun read! I love the birds-as-cocktails conceit, the connections the author finds between them, and the ever-present wordplay—which goes well beyond the book's title to drinks like the Sidecardinal and the Compulsive Lyrebird. The cocktail recipes are original, interesting, and well within reach for readers to mix at home. You *will* learn a lot of great bird facts, but *Make Myna Double* is such a breezy book you may not even realize how much it's teaching you. A clever, cheeky contribution to ornithologist-mixologist relations—and undoubtedly a great gift for bird lovers and cocktail lovers alike!"

—**Brian D. Hoefling**, author of *Dryads: Spirits of the Trees* and *The Cocktail Seminars*

"Katie's new cocktail book is as delightful as it is clever. Effortlessly weaving the study of birds, geography, and history, Katie devises the perfect cocktail with a twist. Easy-to-prepare recipes, with spirits and mixer alternatives, add another layer to this humorous, unique, playful, and beautifully illustrated scientific cocktail adventure."

—**Diane Lapis**, author of *Cocktails Across America*

Make Myna Double

Make Myna Double

Cocktails for Bird Lovers

Written and Illustrated By

Katie Stryjewski

MIAMI

Published by Mango Publishing, a division of Mango Publishing Group, Inc.

Cover Design: Elina Diaz
Cover Photo/Illustration: Katie Stryjewski
Layout & Design: Elina Diaz

For permission requests, please contact the publisher at:
Mango Publishing Group
5966 South Dixie Highway, Suite 300
Miami, FL 33143
info@mango.bz

For special orders, quantity sales, course adoptions and corporate sales, please email the publisher at sales@mango.bz. For trade and wholesale sales, please contact Ingram Publisher Services at customer.service@ingramcontent.com or +1.800.509.4887.

Make Myna Double: Cocktails for Bird Lovers

Library of Congress Cataloging-in-Publication number: 2025938538
ISBN: (print) 978-1-68481-797-9, (ebook) 978-1-68481-798-6
BISAC category code: CKB006000, COOKING / Beverages / Alcoholic / Bartending & Cocktails

For my mom, who would have loved this book.

Table of Contents

Introduction

Author's Note

At first glance, I suppose birds and cocktails don't seem to have much to do with each other. In fact, they don't have much to do with each other at second or third glance either. Birding is enjoyed out in nature; cocktails, rarely. Birders get up early; bartenders stay up late. Birding is free to do (at least until you get *really* into it); cocktails, whether purchased at a bar or made at home, can be expensive. Birding is good for your health; drinking cocktails is decidedly not.

And yet, if you are holding this book, you are probably already interested in one of these things and at least curious about the other. I've found a lot of overlap between people who enjoy birds and those who enjoy cocktails. I personally take this to the extreme—they are two of my favorite things, and they have each kept me employed at one or another point in my life. The existence of this book makes a lot more sense when you understand where I'm coming from. So let me briefly answer the question I am frequently asked at dinner parties: how does an ornithologist end up posting cocktails on Instagram for a living?

I've loved birds since I was a kid. When I was in seventh grade, I decided that I wanted to be an ornithologist. I wanted to travel to tropical locations to study exotic birds and discover new species. I did my undergraduate degree at the nearby Louisiana State University, conveniently home to one of the best ornithology departments in the country. I wrote my undergraduate thesis on Jamaican streamertail hummingbirds (see the Doctor Bird cocktail, pg. 81), and I spent my last summer at LSU on a collecting expedition in Peru. I earned my PhD from Boston University, where my dissertation focused on a fascinating group of finch species from Australia and New Guinea (munias in the genus *Lonchura*). I made several trips to Papua New Guinea and one to Australia. I basically fulfilled my childhood dream.

Early in my graduate studies, I won a coveted fellowship, and my husband and I went out to celebrate at a restaurant called Eastern Standard. I was absolutely enchanted by the cocktails there. They contained ingredients I'd never even heard of, much less tasted. They were made with such craft and attention to detail. Each one had a story. People talk about their "spark bird," but I didn't have one. I had a spark cocktail. It was an Aviation.

Of course, these kinds of fancy drinks were pretty expensive, and I was a poor graduate student. So instead of going out, I learned how to make drinks at home. I found the recipes online and began a collection of spirits and liqueurs. I loved learning the history behind the classic cocktails. Since I was always using blogs to find information and recipes, I thought it would be fun to start one as I worked my way through the world of cocktails. And thus, Garnish was born (www.garnishblog.com). I got on Instagram, where my follower count steadily rose. I started getting sent PR mailers from liquor companies, and occasionally a brand would pay me to post about their products.

Around this time, I was completing a post-doc at Harvard and struggling with the idea of looking for a job in academia (for reasons I won't get into here). I took a year off when I had my son, and during that time I managed to make a decent amount of money creating recipes, taking photos, and posting sponsored content to Instagram. I realized that this was something I could actually do for a living. I became a full-time cocktail content creator.

I do miss working with birds, and they often inspire my cocktail names or ideas. I don't remember exactly when the idea for this book came about, but I started collecting birdy cocktail puns and the project grew from there. As you'll see, these pages contain more than just drink recipes and bird illustrations. Work your way through *Make Myna Double* and you'll get a crash course in both home bartending and all my favorite parts of ornithology: evolutionary biology, behavioral ecology, taxonomy, and even genomics. And—unlike my BIO 107 students at BU—you'll be able to drink during class.

After all, perhaps the combination of birdwatching and cocktail-making is not so strange. Both of these hobbies can be extremely fun and rewarding. They are rich with history and community, full of skills to learn and rare finds to hunt down. Most of all, they both encourage us to slow down, temporarily escape the frantic pace of modern life, and take the time to deeply enjoy something.

So I hope you enjoy reading and using this book. I hope you learn something new, make something new, and laugh a little too. Cheers!

Notes on Organization, Nomenclature, and References

In this book, you will find 65 original cocktail and mocktail recipes, each inspired by a species of bird. Many are based on classic drinks. The recipes are organized alphabetically by base spirit (gin, rum, tequila, vodka, and whiskey), with other bases (e.g. Cognac, pisco, sherry) and nonalcoholic recipes at the end. If you are interested in a specific bird, cocktail, or ingredient, check the index. I've swapped the base spirit in some of the classic recipes, so they may not be where you expect.

Bird taxonomy is constantly changing as we learn new information, and not everyone is in agreement as to how some groups and species should be categorized. For the sake of this book, I will be using the taxonomy of the International Ornithological Congress World Bird List as of 2025 (version 15.1). Scientific names are given as *Genus species*, and the genus may be shortened to its first initial after its first appearance in the text (*G. species*). When referring to more than one member of a genus, "*Genus* spp." is used.

I will not cite references for more general and established scientific facts, but I will try to provide sources whenever I discuss recent research or specific figures.

And for all you listers out there—go to the end of the book for a checklist, so you can tick off all the cocktails you make!

Pronunciation Guide

Birds are divided into orders, which are then divided into families. Bird orders all end with *-iformes*, pronounced "if-orm-ees." Families end in *-idae*, usually pronounced "id-ee." If you see two i's next to each other, you pronounce them both—the first as "ee" and the second as the i in "if," with the first stressed (for example, the family of loons, Gaviidae, is pronounced "gav-EE-id-ee").

Field Guide to Tools, Glassware, and Ingredients

Bar Tools

1. **Jigger.** ½–2 oz. Used for measuring liquids. Most have two sides with different volume measures, sometimes with internal markers for smaller volumes. Usually made of stainless steel. **(a) "Japanese" jigger. (b) "Leopold" jigger.**

2. **Shaker.** 17–28 oz. Used to chill, dilute, and mix drinks that contain juices. Typically made of stainless steel, though some have glass, plastic, or rubber components. **(a) Cobbler Shaker:** the "quintessential" cocktail shaker, made up of a tin, a cap, and a lid with a built-in strainer. **(b) Boston Shaker:** made up of two metal tins that fit together at an angle. Requires a separate strainer to pour the cocktail into the glass. Preferred by professional bartenders because the tins do not get stuck. It also gives the cocktail twice as much distance to travel with each shake.

3. **Mixing Glass.** 17–28 oz. For stirring spirit-forward cocktails. Usually made of glass, with a small spout.

4. **Barspoon.** A long, spiraled spoon used to stir a cocktail and as a unit of measurement. One barspoon = 1 tsp.

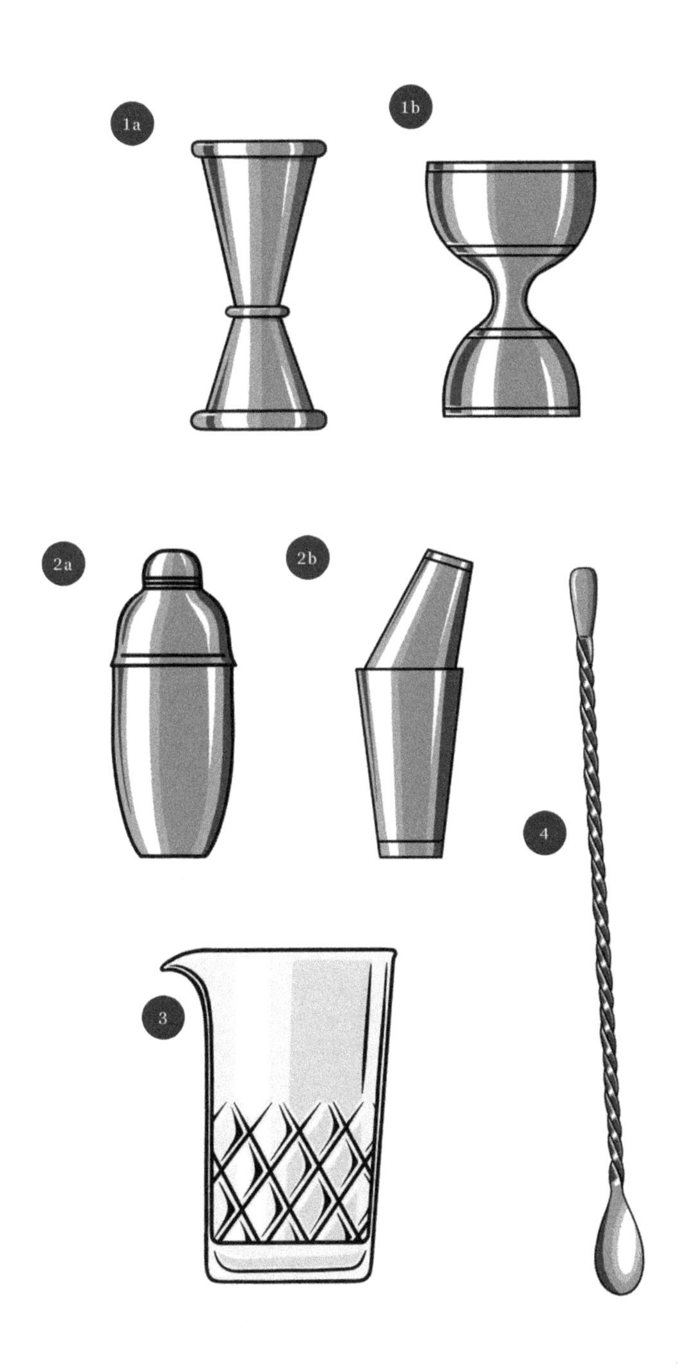
1a
1b
2a
2b
3
4

5. **Strainer.** A tool used to pour cocktails out of a mixing glass or shaker while preventing ice and other solids from falling into the glass. **(a) Hawthorne Strainer.** A flat metal strainer fitted with a spring that fits into the shaking tin or mixing glass. **(b) Julep Strainer.** An old-fashioned style of strainer that looks like a large metal spoon with holes. Traditionally used to sip icy drinks like the mint julep before straws were commonplace, it can also be used to strain cocktails. **(c) Fine Strainer.** A mesh strainer that is used with a Hawthorne or Julep strainer to catch very fine bits of fruit, herbs, and ice that would otherwise slip through. It is held above the glass as the cocktail is strained through it. **(d)** A hybrid Hawthorne/fine strainer known as the **Kilpatrick Strainer.**

6. **Muddler.** A blunt tool used to press fruit or herbs, releasing juice and aromatic oils.

7. **Cocktail Pick.** A sharp metal pick used for spearing garnishes.

8. **Swizzle Stick.** A long wooden stick with short prongs at one end, used for mixing drinks in a tall glass. Drinks mixed this way are usually called "swizzles." Traditional Caribbean swizzle sticks are made from a branch of the "swizzle stick tree," *Quararibea turbinata.*

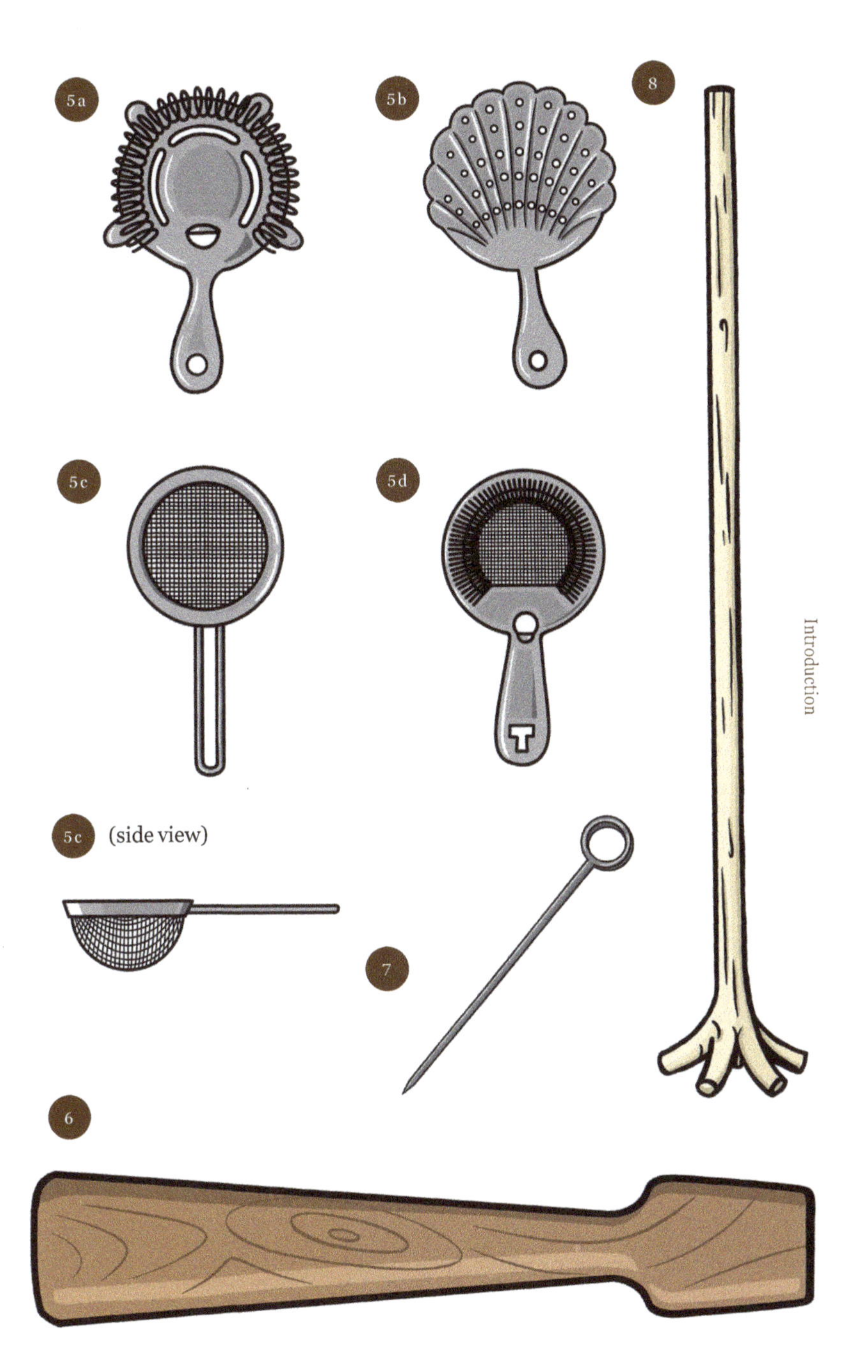
5a
5b
8
5c
5d
5c (side view)
7
6

Glassware

1. **Rocks Glass.** 6–8 oz. Also called an Old-Fashioned glass, lowball glass, or tumbler. A low glass used to serve drinks on ice. Can also be used for "down" or "neat" serves without ice. **Double Old-Fashioned Glass,** 10–12 oz.

2. **Highball Glass.** 8–12 oz. A tall glass used for drinks with a carbonated component that are served over ice. **Collins Glass:** 10–12 oz. While the terms are often used interchangeably, a Collins glass is taller and narrower than a highball glass, and holds more volume.

3. **Cocktail Glass.** Also called a Martini glass. A stemmed glass with a bowl shaped like an inverted cone. While this glass was popular for most of the twentieth century, many craft cocktail bars now opt for coupe glasses (number 4) instead, as they are less prone to spill.

4. **Coupe Glass.** A stemmed glass with a rounded bowl.

5. **Nick and Nora Glass.** A stemmed glass with a rounded bowl that is smaller and narrower than that of a coupe. Commonly used to serve spirit-forward cocktails. Popular in the first half of the twentieth century, this smaller glass was resurrected and given its name by bartender Dale DeGroff in the late 1980s.

6. **Champagne Flute.** A tall, narrow stemmed glass that is slightly tapered at the top. Used to serve champagne and cocktails made with sparkling wine.

1
2
3
4
5
6

7. **Wine Glass.** A stemmed glass for serving wine that is often used for spritzes or Spanish-style gin and tonics.

8. **Brandy Snifter.** A low glass with a wide, rounded base that tapers at the top to concentrate aromas. Traditionally used for brandy. Available in a variety of sizes.

9. **Tiki Mug.** A ceramic cup with a tropical- or Polynesian-inspired shape.

10. **Footed Pilsner Glass.** A tall, stemless, sculptural glass originally intended for beer.

11. **Hurricane Glass.** A tall glass that narrows in the center, with a short stem. Named for the Hurricane cocktail served at Pat O'Brien's in New Orleans, and for its shape, which is similar to that of a hurricane lamp.

12. **Irish Coffee Mug.** A glass mug used for hot drinks.

13. **Moscow Mule Mug.** A copper mug created to serve the Moscow Mule cocktail. Primarily used for this and other cocktails made with ginger beer (commonly called mules or bucks).

7
8
9
10
11
12
13

Ingredients

Main Base Spirits

1. **Gin.** A spirit flavored with herbs and botanicals, particularly juniper. **Indigo morph.** Indigo gins are infused with butterfly pea flower. They turn pink when acidic ingredients are added.

2. **Rum.** A spirit distilled from sugarcane products, usually molasses. Can be categorized many ways. For the purposes of this book: **Aged rum** is a brown rum that has been barrel-aged. **White rum** is usually aged as well, but has been filtered to remove color. **Black rum** (sometimes called blackstrap rum) is very dark and often heavily colored and sweetened. Its thick texture is an important factor in recipes that call for it specifically.

3. **Tequila.** A spirit distilled from blue agave in specific parts of Jalisco, Mexico. **Blanco tequila** is aged for less than two months and is colorless, with bright, more vegetal flavors. **Reposado tequila** is aged for up to one year and has a pale golden color and flavors of vanilla and caramel. **(a) Mezcal.** While technically referring to any spirit distilled from agave, mezcal in the United States is usually distinctly smoky.

4. **Vodka.** A colorless, flavorless spirit.

5. **Whiskey.** A spirit distilled from grains such as corn, rye, barley, or wheat that is aged in barrels. **Bourbon** is an American whiskey made from at least 51 percent corn. **Rye** is an American whiskey made from at least 51 percent rye. **Scotch** is a Scottish whiskey traditionally made from malted barley.

1
Batavia-Arrack
van Oosten
JAVA SEA
INDIAN OCEAN
2
novo fogo
silver cachaça
3
1840
COGNAC
PIERRE
FERRAND
1840

4
QUEBRANTA
Producto de Perú
SUYO
PISCO
VALLE DE MALA
Single Origin
PISCO PURO
QUEBRANTA
5
TRÄ KÁL
TRÄ·KÁL
750 ml

Other Base Spirits

1. **Batavia Arrack.** An Indonesian spirit made from sugarcane and red rice. A classic ingredient in historical punches.

2. **Cachaça.** A Brazilian spirit made from sugarcane juice.

3. **Cognac.** A subspecies of brandy endemic to the Cognac region of France. Brandy is a spirit distilled from grapes.

4. **Pisco.** A spirit distilled from grapes that is native to Peru and Chile.

5. **Trakal.** A spirit native to Chilean Patagonia distilled from apples, pears, and crabapples.

1
2a
2b
ABSENTE
ABSINTHE REFINED
110
WORMWOOD
CAMPARI
APEROL
APERITIVO
1919
2c
3a
3b
LUXARDO
BITTER
BIANCO
ANGOSTURA
aromatic bitters
44.7% alc/vol
4 fl.oz. (118 mL)
FEE BROTHERS
1864
West Indian
ORANGE
BITTERS
5 FL OZ (150 ML)

Modifiers

Liqueurs, fortified wines, and other alcoholic cocktail ingredients are often referred to as modifiers. They are added to the base spirit to modify the flavor of the drink. If you don't see a particular one listed here, it will be explained under the recipe that calls for it.

1. **Absinthe.** A spirit flavored with wormwood, anise, fennel, and other botanicals with a licorice-like flavor. Alternatives include Herbsaint and pastis.

2. **Amaro.** A bittersweet liqueur, usually made in Italy. **Aperitif** amari are more bitter than sweet and are meant to be consumed before a meal. The most famous are **(a) Campari** and **(b) Aperol.** Others used in this book include **(c) Luxardo Bitter Bianco,** Martini Fiero, and Suze. **Digestif** amari are traditionally consumed after a meal, and are strongly flavored with herbs and botanicals. Varieties used in this book include Cynar, Fernet Vallet, Jaegermeister, Nonino, and Ramazzotti.

3. **Bitters.** Concentrated extracts of bitter botanicals and other flavors. Think of them as the salt and pepper of the bar world. **(a) Angostura bitters** from Trinidad are by far the most common; **(b) Orange bitters** also appear in many recipes. Other varieties used in this book include grapefruit, eucalyptus, lavender, Peychaud's, cherry bark vanilla, and yuzu.

4a
1849
COINTREAU
LIQUEUR
4b
PIERRE FERRAND
Dry CURAÇAO
PF
4c
Bols
5
LUSTAU
Amontillado
LOS ARCOS
6a
DOLIN
DOLIN
VERMOUTH
6b
ANTICA FORMULA
1786
6c
DOLIN
DOLIN
VERMOUTH
6d
LILLET
LILLET BLANC

4. **Orange Liqueur.** A sweetened spirit infused with orange flavor. **(a) Triple Sec** is a neutral spirit infused with orange peel, originating in France. Cointreau is my triple sec of choice. **(b) Curaçao** is a brandy base infused with dried orange peel and sometimes other herbs and spices, originating in the country of Curaçao. Look for Pierre Ferrand Dry Curaçao. **(c) Blue morph.** Blue curaçao is essentially just orange curaçao with food coloring.

5. **Sherry.** A fortified wine from Spain. Sherry should be stored in the refrigerator to preserve its quality and flavor. Two varieties are used in this book: **amontillado,** which is dry and nutty, and **oloroso,** which is richer and darker.

6. **Vermouth.** A fortified wine infused with botanicals, traditionally including wormwood (*wermut* is the German word for wormwood). Like sherry, vermouth should be stored in the refrigerator to preserve its quality and flavor. **(a) Dry Vermouth.** A white vermouth that is light, herbaceous, and dry. **(b) Sweet Vermouth.** A red vermouth that has been sweetened. It has flavors of dark fruits and baking spices. **(c) Blanc** or **Bianco Vermouth.** This "white" vermouth is lightly sweetened and between the other two in body and flavor. **(d) Lillet.** A floral French aperitif wine often used like a vermouth. **Lillet Rosé** is a fruity, rosé subspecies of Lillet.

7a
St. Elizabeth Allspice Dram
7b
DISARONNO
DISARONNO
ILLVA · SARONNO
7c
GIFFARD
BANANE DU BRESIL
7d
EXPLORING THE WORLD SINCE 1818
HEERING
CHERRY LIQUEUR
7e
CRÈME DE CACAO
TEMPUS FUGIT SPIRITS
7f
MÛRE
1885
GIFFARD
MÛRE
7g
MR·BLACK
LIQUEUR

7. **Other Liqueurs.** Liqueurs are spirits that have been flavored and sweetened.

    (a) Allspice Dram. A liqueur flavored with allspice berries, giving it flavors of cinnamon, clove, and nutmeg. Also called pimento dram. St. Elizabeth is the most common brand.

    (b) Amaretto. A sweet almond-flavored liqueur made from apricot kernels, peach stones, and/or almonds.

    (c) Banana Liqueur. A sweet, banana-flavored liqueur. Look for Giffard Banane du Bresil or Tempus Fugit's Crème de Banane.

    (d) Cherry Heering. A dark, rich cherry liqueur from Denmark popularly used in the Singapore Sling and Blood and Sand cocktails.

    (e) Crème de Cacao. Chocolate liqueur (which, despite the name, is not made with cream). Tempus Fugit Crème de Cacao is excellent.

    (f) Crème de Mûre. A thick and sweet blackberry liqueur.

    (g) Coffee Liqueur. A liqueur containing coffee or espresso. I like Mr. Black Cold Brew Coffee Liqueur and St. George NOLA Coffee Liqueur, both of which have intense coffee flavor without being too sweet.

ST-GERMAIN

7i

RUM CHATA
Horchata Con Ron

LUXARDO
GIROLAMO LUXARDO

7k

CHINOLA
PASSION FRUIT
LIQUEUR

(h) Elderflower Liqueur. A sweet, floral liqueur flavored with elderflowers. St. Germain is the best-known brand.

(i) Horchata Liqueur. Horchata is a creamy beverage usually made from rice and flavored with vanilla and cinnamon. Expect to find those flavors in cream liqueurs such as RumChata and Ricura.

(j) Maraschino Liqueur. A liqueur made in Italy and Croatia from Maraska cherries that is an ingredient in many classic cocktails. It is lighter and more candy-sweet than Cherry Heering (d).

(k) Passion Fruit Liqueur. A tart and tropical addition to cocktails. Chinola is a delicious example.

Syrups

Many cocktails contain some kind of syrup as a sweetener. The most common is simple syrup, which is aptly named—this one-to-one mix of sugar and water is easy to make. Flavored syrups are only a little bit harder. The four syrups below are used multiple times in this book. The recipes for the remaining syrups are included with the cocktail recipes where they are used. If a recipe is not included, then I recommend purchasing that syrup (e.g. orgeat, falernum, violet syrup, and lavender syrup—though all can be made at home if you're up for it).

SIMPLE SYRUP

1 cup sugar
1 cup water

Combine sugar and water in a saucepan over medium heat. Bring to a simmer, stirring frequently until sugar is dissolved. Let cool completely before using. Simple syrup should last for several weeks in the refrigerator; throw it out if you notice any cloudiness.

HONEY SYRUP

Follow the recipe for simple syrup but replace the sugar with honey.

GINGER SYRUP

1 cup sugar
1 cup water
1 three-inch piece of fresh ginger, peeled and sliced

Combine sugar and water in a saucepan over medium-low heat. Bring to a simmer, stirring until sugar is dissolved. Add the sliced ginger and stir. Allow the mixture to simmer for a minute or two, then remove from the heat and cover. Let sit until fully cooled, then strain out the ginger pieces. Should last in the refrigerator for at least two weeks.

Ingredient Note: If you're making another recipe that requires peeled ginger, save the scraps and peels—you can also infuse the syrup with those for a more sustainable version of this recipe.

PASSION FRUIT SYRUP

1 cup passion fruit puree
1 cup sugar

In a saucepan, combine passion fruit puree and sugar. Bring to a simmer, stirring until the sugar has dissolved. Let cool completely, then fine-strain to remove any solids. Store in the refrigerator for up to two weeks.

Ingredient Note: If you have trouble finding passion fruit puree, check your grocery store's frozen fruit section.

Techniques

Measuring

Cocktail ingredients in the United States are typically measured in ounces using a jigger. Jiggers come in sizes from ½ to 2 oz. If you don't have one, you can use a tablespoon (1 tbsp = ½ oz). Fill the appropriately sized jigger to the very top for an accurate measurement. You may have to estimate if the measure is less than the full volume of the jigger (e.g. measuring ½ oz in a 1-oz jigger). Other measures you may encounter are 1 barspoon (equal to 1 tsp), 1 dash (a single shake of a bitters bottle, or ⅛ tsp), and a pinch (used for solids, also ⅛ tsp).

You may encounter recipes with a mixer such as club soda or sparkling wine added "to top." This means to fill the remainder of the glass with the mixer. However, this assumes that you are using a standard-size glass, so keep that in mind when pouring.

The ingredients in cocktail recipes are typically listed with the alcoholic ingredients first, in order of volume, followed by the nonalcoholic ingredients. When building your cocktail, it is common to add the ingredients from the bottom up. That way, if you make an error, you can start over without wasting your most expensive ingredients. This is more of an issue in a high-volume cocktail bar where small pours add up, but it never hurts to be frugal.

Mixing

Almost all cocktails are stirred or shaken with ice before being served. This does three things: it combines the ingredients, it chills the cocktail, and it dilutes the cocktail. Diluting your drink may not sound like a good thing, but it's actually an important step in properly balancing it. Adding water releases esters and other volatile compounds that give spirits their flavor.

So, when do you a shake a drink and when do you stir? Cocktails are shaken when they contain citrus juice. Juice has a very different density from alcohol, and shaking helps to fully incorporate the two. It also pleasantly aerates the drink and improves its texture. Stirring, on the other hand, is used for spirit-forward cocktails where all the ingredients have similar densities and a smooth texture is preferred.

Shaking

For a shaken cocktail, add all of your ingredients to the tin of your shaker. If you're using a Boston shaker, you can build the drink in either tin. Be careful not to add any carbonated ingredients at this point; mixers like club soda or sparkling wine are always added directly to the glass. Once all of your ingredients have been added, fill about ¾ of the shaker with ice. Don't use crushed or pebble ice to shake a drink unless the recipe calls for it, as this can over-dilute the cocktail. Ideally, cubes for shaking are around 1" square. The ice from your home freezer should work well.

Once you've added the ice, seal the shaker. For a cobbler shaker, this means putting on the lid and cap. Make sure to put the lid on straight, as placing it at an angle can cause it to get stuck. For a Boston shaker, on the other hand, you want to place the smaller tin inside the larger one at an angle and give it a firm rap on the base to create a seal. Then shake your drink for about ten seconds, until the outside of the shaker begins to get frosty and the ice inside starts to break up.

Stirring

To stir a cocktail, add all of your ingredients to a mixing glass. Fill about ¾ of the mixing glass with ice. Using a barspoon, stir the cocktail for around thirty seconds by twirling the spoon along the inside walls of the glass.

Straining

Once your cocktail is mixed and chilled, it's time to strain it into your glass. If you're using a cobbler shaker, remove the small cap and pour. For a Boston shaker, you'll need to separate the tins. If you have trouble, try squeezing the sides of the larger tin or giving the shaker a firm rap with your hand a quarter of the way around from where the two tins meet.

Insert your strainer into the shaker tin or mixing glass. Hawthorne strainers sit with the spring just inside the rim of the glass, while julep strainers fit into the glass at an angle with their handle resting on the edge. Pour through the strainer into your glass. For cocktails with muddled fruit or herbs, or if you want to prevent any ice chips from falling into the glass, hold a fine strainer over the glass as you pour.

Muddling

Muddling is a great way to incorporate the flavors of fruit and herbs into your cocktail. Place whatever needs to be muddled in the bottom of your shaker (or directly into your glass if the recipe calls for it). Recipes often suggest adding at least one liquid ingredient, usually syrup, to the solids that are going to be muddled. This gives all those oils and flavors somewhere to go as you extract them.

To muddle, gently press down on the ingredients with your muddler, twisting it slightly with your wrist. When muddling herbs, you simply want to press and bruise them to expel oils; try not to tear them into pieces, which can release bitter flavors. You can be less gentle when muddling fruit, being sure to press out all its juice.

Fine-straining is usually recommended after muddling, as small pieces of solids can easily slip through a standard strainer and into your drink.

Garnishing with a Twist

If a recipe says to garnish a cocktail with a citrus twist, this doesn't mean just dropping one into the glass or setting it on the rim. When done right, citrus twists are very functional garnishes that add a great deal of flavor and aroma to a drink. To garnish with a citrus twist, use a knife or peeler to remove a section of peel from your fruit. Try to get only the outer portion of the peel, leaving the white pith behind. Position the peel over your drink, holding it in two hands with the outer side facing the liquid. Pinching it between your thumbs and forefingers, gently squeeze to fold the peel in half longways so that the center bends toward the drink. As you do so, you should see a burst of citrus oil hit the surface. Repeat this a couple of times and then run the outside of the peel along the rim of the glass before dropping it into the drink or discarding it.

Native Species

There are cocktails out there with all sorts of unusual names: Monkey Gland, Fuzzy Navel, Suffering Bastard... the list goes on. So it's a bit surprising that there aren't *that* many classic cocktails named after birds. Only a choice few avian-themed recipes have withstood the test of time. But as they are so few, they are certainly worth knowing. Here are five classic cocktail recipes named for birds.

Cardinale

This cocktail is a variation of the Negroni that calls for dry vermouth instead of the Negroni's usual sweet. It hails from Rome, where the story goes that bartender Giovanni Raimondo of the Excelsior Hotel named it in honor of a Catholic cardinal who was visiting for the 1950 Jubilee. If you find it too bitter or intense, try using Aperol in place of Campari (a variation called a Contessa).

1 oz gin
1 oz dry vermouth
1 oz Campari

Combine all ingredients in a mixing glass with ice and stir until chilled. Strain into a rocks glass filled with ice. Garnish with an orange twist.

Jungle Bird

A Tiki favorite, the Jungle Bird was created at the Kuala Lumpur Hilton's Aviary Bar by Jeffrey Ong in the 1970s. It's often served in a ceramic mug shaped like a macaw.

1 ½ oz black rum
¾ oz Campari
2 oz pineapple juice
½ oz lime juice
½ oz simple syrup

Combine all ingredients in a shaker with ice and shake until chilled. Strain into a Tiki mug filled with ice. Garnish with a wedge of pineapple and several pineapple fronds.

Paloma

Paloma means "dove" in Spanish. The origins of this cocktail are murky, but we do know that it came from Mexico and that it has been around since at least the 1950s. The name may have come from a popular Spanish folk song from the 1860s, "La Paloma." The version below uses fresh grapefruit juice, but you can also mix tequila with grapefruit soda and a squeeze of lime for a quick and easy cocktail.

2 oz blanco tequila
2 oz pink grapefruit juice
½ oz lime juice
½ oz simple syrup
2 oz club soda, to top

Rim a highball glass with salt and fill it with ice. Combine tequila, grapefruit juice, lime juice, and simple syrup in a shaker. Add ice and shake until chilled. Strain into the prepared glass and top with club soda. Garnish with a slice of grapefruit.

Stork Club

The Stork Club was a famous Manhattan bar that was open from 1929 until 1965. Since Prohibition didn't end until 1933, it operated illegally for its first few years. As such, it might not surprise you to know that the club's owner, Sherman Billingsley, was a former bootlegger who was involved with the mob. Billingsley claimed he couldn't remember why he called his bar the Stork Club, but the name was passed on to the club's signature drink. The original recipe called for only a dash of lime juice and orange liqueur, but here I opt for a slightly more substantial ¼ oz to make the drink more appealing to modern palates.

Ingredient Note: Orange juice pops up in many Prohibition-era recipes. Try squeezing your own—it really does make a difference in taste and texture.

1 ½ oz gin (preferably Old Tom)
¼ oz orange liqueur
¾ oz orange juice
¼ oz lime juice
1 dash Angostura bitters

Combine all ingredients in a shaker with ice and shake until chilled. Strain into a chilled coupe. Garnish with an orange twist.

Ingredient Note: Old Tom is an older style of gin that is sweeter than today's typical London Dry.

Yellow Bird

Like the Paloma, the Yellow Bird probably got its name from a song. "Choucoune," a lyrical poem written in Haitian Creole in 1883, became a popular song when it was put to music ten years later. When the song was translated into English in the 1950s, the lyric *"ti zwazo"* ("little bird") was translated as "yellow bird," and this became the title. Like the song, the cocktail is certainly Caribbean in origin, but recipes for it can vary greatly. This is the one I like best.

¾ oz aged rum
¾ oz white rum
½ oz banana liqueur
½ oz Galliano
1 ½ oz orange juice
1 oz pineapple juice
¼ oz lime juice

Combine all ingredients in a shaker with ice and shake until chilled. Strain into a highball glass filled with ice. Garnish with a slice of orange and a cherry.

Ingredient Note: Galliano is a sweet, herbal Italian liqueur that was popular in the mid-twentieth century. Its flavor profile includes anise, vanilla, and cinnamon. If you don't have Galliano, substitute another anise-flavored liqueur like absinthe, pastis, or sambuca. Using ¼ oz of one of these plus ¼ oz Licor 43 would be even closer.

Recipes

Gin

Bee-Eater's Knees

When flipping through a field guide to the birds of Europe, you don't really see a lot of color. Most of the continent's birds sport plumage in drab shades of brown and grey. But the European bee-eater (*Merops apiaster*) is a brilliant exception. This vividly colored bird is highly social, nesting in groups of burrows dug in the sides of sandy riverbanks. As its name would suggest, the European bee-eater eats insects, especially bees. Swallowing something with a stinger can be a risky business, but bee-eaters are pros. Before swallowing a bee, they will strike it repeatedly against a branch to kill or stun it, then rub its abdomen against the side of the branch to remove the stinger and discharge any venom.

The Bee's Knees is a Prohibition-era cocktail made from gin, honey, and lemon juice. The colors in this "long" version (meaning a carbonated mixer has been added) are made to mimic the bee-eater's spectacular plumage. Back in the 1920s, saying something was "the bee's knees" meant it was really great. If you ask me, the Bee-Eater's Knees is even better.

Bee-Eater's Knees

2 oz indigo gin
¾ oz lemon juice
½ oz honey syrup (pg. 37)
2–3 oz Spindrift Pink Lemonade (or other unsweetened lemon seltzer)
½ cup honeydew melon, cut into chunks

Combine lemon juice, honey syrup, and honeydew melon in a shaker and muddle well to release the honeydew juice. Fill the shaker with ice and shake until chilled. Fine-strain into a highball glass filled with ice. Slowly top with the seltzer and then with the gin, creating a layered effect. Garnish with melon balls skewered on a cocktail pick.

Pro Tip: Pour each layer slowly over the back of a spoon and/or along the side of the glass to achieve the best effect. But be sure to stir before you enjoy the cocktail!

Bitter-n

If you've ever been birding in a North American marsh, you may have seen an American bittern (*Botaurus lentiginosus*) in its unique, stick-straight posture, its beak pointing into the air. Or you may *not* have seen it, as this position makes for fantastic camouflage. The bittern instinctively adopts this posture when it feels threatened. The brown streaks on its neck and breast, which run horizontally or diagonally while it hunts, become oriented vertically, mimicking the blades of marsh grass surrounding it. If it stands perfectly still, it is easy to overlook.

Bitterns aren't the only birds who use a specific posture to hide in plain sight. Owls are experts at positioning themselves so as to disappear into tree cavities. Potoos lay their eggs in depressions on top of broken branches or stumps and then assume a position very similar to that of the bittern while they incubate, making them look like the broken-off tip of the branch. They have slits in their eyelids that allow them to keep watch while their massive eyes stay shut, maintaining the illusion.

Bitter flavors play a huge part in craft cocktails, as is evidenced by the hundreds of amari and bitters on the market. A little bit of bitterness in a drink can help marry ingredients together or enhance flavors that were previously getting lost. A *lot* of bitterness can be an acquired taste. If you have acquired it, this is the cocktail for you.

Bitter-n

1 oz gin
1 oz Punt e Mes
1 oz Cynar
½ oz Ramazzotti

Combine all ingredients in a mixing glass with ice and stir until chilled. Strain into a rocks glass over one large ice cube. Garnish with an orange twist.

Ingredient Note: Punt e Mes is an Italian vermouth that is more bitter than most (the name translates to "a point and a half," referring to it being one part sweet and a half part bitter). If you don't have it, substitute sweet vermouth. You can also try replacing the Ramazzotti with another amaro (Montenegro, Lucano, or Averna would all work well) in a pinch.

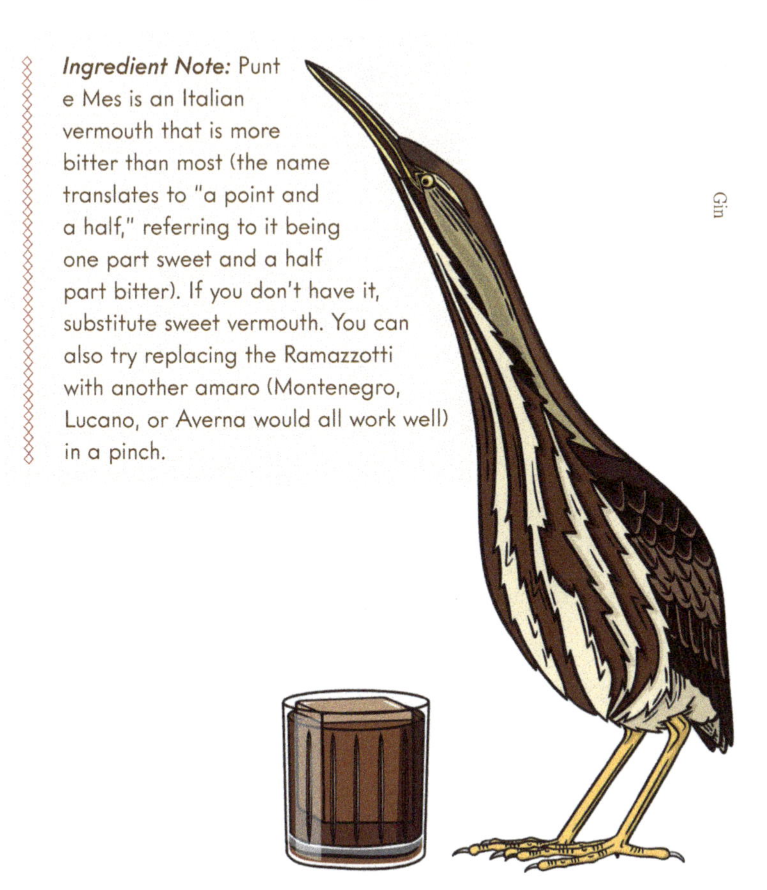

Booby Trap

You can tell a lot about a blue-footed booby (*Sula nebouxii*) from its feet. The vivid color of blue-footed booby feet comes from a combination of carotenoid pigments and collagen structure. As discussed under the Scarlet Rye-bis (pg. 172), birds can't make their own carotenoids and need to get them from their diet. For this reason, carotenoid color is often considered an example of an **honest signal** of the fitness and quality of an individual. In other words, they can't fake it. A booby with very blue feet must be able to find and catch plenty of carotenoid-rich fish to keep that color bright. There is also evidence that carotenoids give birds better resistance to disease and parasites, meaning that boobies with more vivid feet are healthier too.[1] So when blue-footed boobies flaunt that vibrant blue during courtship displays, it's not just for flash; they're really showing you what they've got.

Fans of the Negroni or lesser-known Lucien Gaudin will enjoy this bitter, layered cocktail inspired by the blue-footed booby's most famous feature.

Booby Trap

1 oz gin
1 oz dry vermouth
½ oz Luxardo Bitter Bianco
½ oz blue curaçao

Carefully add blue curaçao to the bottom of a chilled cocktail glass. In a mixing glass, combine gin, dry vermouth, and Luxardo Bitter Bianco. Add ice and stir until well-chilled. Carefully strain this mixture down the side of the cocktail glass to create a layered effect. Garnish with an orange twist (discarded).

Gin & Tit

Boobies, woodcocks, shags...the bird world has its share of suggestive names, but great tits (*Parus major*) are surely the pinnacle. These common European garden birds have a name that could cause some serious misunderstandings for a binocular-toting birder. However, the double entendre is entirely accidental. "Tit" is an Old English word for anything small, and in the case of the great tit and its relatives, it's a shortened version of "titmouse" (which is also a bird, not a mouse, but if we dive too deep into ornithological etymology, we'll be here all day). We have tits all over America too, but at some point during the nineteenth century, our more Puritan forefathers saw fit to rename most of them chickadees after their "chick-a-dee-dee-dee" call. The tufted titmouse is the only one to retain a hint of its true salacious nomenclature.

The Gin & Tit is based on the Gin & *It*, a cocktail that (like the word "chickadee") has its origins all the way back in the 1800s. The "It" does not, as I once thought, refer to some vague *je ne sais quoi*—it's short for "Italian vermouth," which is what we now call sweet vermouth. The Gin & It is essentially a gin Manhattan. There's a reason it hasn't maintained the popularity of its whiskey-based cousin: it's just not very good. But with a little inspiration from the great tit and its European home—specifically, an English country garden—the Gin & Tit can be a lovely, floral, spirit-forward cocktail to sip while, ahem...looking at a pair of great tits.

Gin & Tit

1 ½ oz gin
1 oz Lillet Rosé
½ oz Pimm's No. 1
1 dash lavender bitters

Combine all ingredients in a mixing glass. Add ice and stir until chilled. Strain into a chilled coupe and garnish with a lemon twist.

Ingredient Note: Pimm's No. 1 is a British gin-based liqueur flavored with fruits and spices.

Honeyguide

The greater honeyguide (*Indicator indicator*) is a fascinating bird, and I'm not just saying that because I studied it as part of my PhD research. For one thing, it has a **mutualistic relationship** with humans in Africa. Greater honeyguides will lead hunters to beehives so that they will break them open to harvest honey, giving the honeyguide access to its favorite treats of beeswax and larvae. Honeyguides even recognize and respond to distinct calls made by the people living in their region.[2]

Honeyguides are also **brood parasites,** a reproductive strategy I discuss more under the Piña Koel-ada (pg. 92). They lay their eggs in the nests of other species of birds and force those birds to raise their chicks. This alone would be diabolical enough, but the greater honeyguide doesn't stop there. Once the honeyguide egg hatches, the blind and naked chick will instinctively pierce any other eggs or chicks in the nest with its sharp hooked bill. Before it is even an hour old, it has already brutally murdered all of its adoptive siblings, ensuring that it receives the parents' full attention.

Naturally, these host species do *not* want a honeyguide egg in their nests, and it is to their advantage to be able to recognize a parasite egg and throw it out. So greater honeyguides have evolved to lay eggs that look like the eggs of their hosts.

Female honeyguides (who carry the genes for egg size and shape) remember what kind of nest they were raised in, so they know where to lay their eggs to make sure they will be a good match.[3]

The greater honeyguide certainly understands that you need to break a few eggs to make an omelet—or a flip. Flips are cocktails made with a whole egg shaken into them, and they've been made for hundreds of years. The egg gives them a thick and velvety texture.

Honeyguide

1 ½ oz Barr Hill Tom Cat Gin
½ oz oloroso sherry
½ oz honey syrup (pg. 37)
1 egg

Combine all ingredients in a shaker with ice. Shake very well to chill the cocktail and froth the egg. Fine-strain into a coupe glass and garnish with grated nutmeg.

Ingredient Note: Barr Hill's Tom Cat Gin is a barrel-aged gin made with honey, which makes it particularly perfect for this recipe. You can substitute another barrel-aged gin or try your favorite whiskey.

Maharaja's Kookaburra-Peg

Unless you're a true cocktail geek, you probably don't know the drink this kookaburra-themed recipe is referencing. The Maharaja's Burra-Peg comes from a particularly wonderful cocktail book by Charles H. Baker bearing the unwieldy title *The Gentleman's Companion: Being an Exotic Drinking Book, or Around the World with Jigger, Beaker, and Flask.* Published in 1939, it includes recipes and stories from Baker's travels, which he appropriately refers to in the introduction as "liquid field work." The Maharaja's Burra-Peg is the first of five champagne cocktails Baker lists under the heading "FIVE DELICIOUS CHAMPAGNE OPPORTUNITIES, which Are not to be Ignored." (Somehow, the haphazard capitalization makes it even better.) Baker writes that he encountered this cocktail in Amber, India, where he drank four of them at sunset on Washington's birthday while an expat told him stories about the local Maharaja. It contains a hefty pour of Cognac, an Angostura-soaked sugar cube, plenty of champagne, and a lime twist. *Burra*, Baker explains, is the Hindu word for "big" or "important," and *peg* is British slang for a drink, usually referring to Scotch and soda.

So what makes a Kookaburra-Peg? The laughing kookaburra (*Dacelo novaeguineae*) has become a national symbol of Australia, the only place where it is found. Despite their unique name, kookaburras are actually kingfishers, and the laughing kookaburra is the largest member of the kingfisher family. This cocktail incorporates the quintessentially Australian flavor of eucalyptus into the Burra-Peg. To take the recipe even further Down Under, use an Australian gin (I recommend Four Pillars) and an Australian sparkling wine like Jansz.

Maharaja's Kookaburra-Peg

1 oz gin
1 sugar cube
3 dashes eucalyptus bitters
3 oz sparkling wine

Splash a dash of bitters onto your sugar cube for it to soak up. Add gin and sparkling wine to a champagne flute. Drop in the sugar cube and garnish with a lemon twist.

Ingredient Note: both Bitter Queens and Hella Bitters make eucalyptus bitters.

Ptarmigimlet

Birds are masters of camouflage. True, some species are made to be showy, but most do their best to stay safe from predators by fading into the background. But what if that background changes over time? A white bird adapted to blend in with the snow would stick out like a sore thumb on the summer tundra, and vice versa. So subalpine birds like the willow ptarmigan (*Lagopus lagopus*) have to change too.

All birds shed and regrow their feathers at least once annually, a process called **molting.** The willow ptarmigan molts twice a year, from white to brown in the spring, and from brown to white in the autumn. They don't lose all their feathers at once, of course—molting is usually an extended process, and you'll see birds with patches of different-colored feathers during these transitional times. By changing its plumage in this way, the willow ptarmigan stays camouflaged in its habitat year-round.

The Gimlet is a classic sour that mixes gin with lime juice and simple syrup (or, more traditionally, lime cordial). Gin is defined by the presence of juniper, which already gives it a resinous, piney flavor reminiscent of a boreal forest. This Ptarmigimlet uses St. George Spirits's Terroir Gin, a distinctive, pine-forward gin meant to evoke evergreen forests, hillside thickets, and damp earth. Fresh rosemary enhances these flavors. The combination will make you feel like you're standing at the edge of the forest, looking out over the tundra, wondering if there are any ptarmigans hiding out there.

Ptarmigimlet

2 oz St. George Terroir Gin
¾ oz lime juice
¾ oz simple syrup
1 large sprig of fresh rosemary

Combine all ingredients, including rosemary, in a shaker with ice. Shake until well-chilled. Fine-strain into a chilled coupe.

Ingredient Note: If you don't have St. George Terroir Gin, go with a juniper-forward gin such as Sipsmith VJOP or Junipero, or look for pine- and spruce-heavy gins from your local distilleries. You can also add some alpine flavor to your favorite gin by mixing ¼ cup spruce tips with 1 cup gin and allowing it to infuse overnight.

Purple Martini

Many people love birds, and when it comes to the purple martin (*Progne subis*), the feeling is mutual. Purple martins are **synanthropic,** which means that they have developed—and come to depend on—a beneficial relationship with humans. Native Americans would hang hollowed-out gourds to encourage these insectivores to nest near their homes as a form of pest control, a trick that European colonists began to imitate. When species like European starlings and house sparrows were introduced to North America and started hogging the purple martins' nesting sites, their population began to decline, and they became dependent on these artificial nest cavities. Today, many bird enthusiasts set up martin houses in their backyards, and it continues to be an important practice to protect this beautiful bird.

While you sit on your patio on a summer evening, enjoying the view of purple martins going in and out of their new homes, consider sipping on an appropriate libation: the Purple Martini. The vivid color of this drink comes from indigo gin made with butterfly pea flower, a unique ingredient that can impart its purple shade to tea and spirits. The color will become pinker when anything acidic is added.

Purple Martini

2 oz indigo gin
½ oz dry vermouth
½ oz blanc vermouth
1 dash grapefruit bitters

Combine all ingredients in a mixing glass with ice. Stir until chilled. Strain into a chilled cocktail glass or coupe. Garnish with a grapefruit twist.

Ingredient Note: There are several gins made with butterfly pea flower on the market, including Empress 1908, The Illusionist, and McQueen and the Violet Fog Indigo Edition. Alternatively, you can use a butterfly pea flower tincture or infuse your own gin (or any clear spirit) with dried butterfly pea flowers. Find them online and at specialty herb and tea shops.

Rum

Bananaquit

How do birds get their spectacular colors? It's in their DNA.

Scientists have long wondered what genes could be responsible for the wide array of plumage colors and patterns that we see in birds. As DNA sequencing becomes cheaper and easier, we are getting more and more genetic data from all sorts of species. But sequencing is only the first step in understanding genomes, because knowing the sequence of a gene doesn't necessarily tell you what it *does*. To figure this out in wild birds, scientists sequence their DNA and look for connections between individual genes and traits like bill shape or plumage color. The first wild bird for which this was ever done successfully was the bananaquit (*Coereba flaveola*).

Bananaquits are small nectar-eating birds that are common throughout Central and South America and the Caribbean. They are striking little birds with yellow underparts, a conspicuous black cap, and a white eye stripe—except on a handful of Caribbean islands, where all-black bananaquits are found alongside the more colorful members of their species.

By studying bananaquit feathers, ornithologists determined that the black color is caused by melanin deposits. A team of researchers sequenced a gene in the bananaquits that is known to regulate the synthesis of melanin, called the melanocortin-1 receptor (MC1R).

Sure enough, they found a single mutation in the MC1R gene that was shared by all the black bananaquits.[4] Since then, MC1R mutations have been found in other wild bird species with melanic forms, including Snow Geese and Parasitic Jaegers,[5] and a number of other color genes have been identified.

What else should one put in a Bananaquit cocktail but banana? The bird gets its name from its bright yellow color. This drink's hue is more subdued, but its flavor is anything but. It's a creamy, spiced blend of delicious tropical flavors.

Bananaquit

- 1 ½ oz aged rum
- ¾ oz banana liqueur
- ¼ oz allspice dram
- ¾ oz lime juice
- ¾ oz cream of coconut

Combine all ingredients in a shaker with ice and shake until chilled. Strain into a footed pilsner glass filled with crushed ice. Garnish with a bouquet of fresh mint, a slice of banana, and an edible flower.

Blue Hawaiian Honeycreeper

Chains of islands, otherwise known as archipelagos, are excellent places for diverse groups of species to evolve, and the Hawaiian archipelago is no exception. It is home to the spectacular Hawaiian honeycreepers. These relatives of the cardueline finches have a wide array of plumage colors and bill shapes, as well as delightful names like ʻŌʻū and ʻAkiapōlāʻau (*Hemignathus wilsoni*, pictured).

Most of them are also extinct.

Sadly, the impact of human settlement on Hawaii has been catastrophic for its bird life. The first Polynesian settlers caused many extinctions through hunting, habitat destruction, and the introduction of domestic animals. European colonists accelerated this by bringing avian malaria to the island, a disease against which the honeycreepers have no defense. Of the fifty known species of honeycreeper, only seventeen remain today.[6] Four of these have fewer than 2,000 individuals left. The ʻAkikiki (*Oreomystis bairdi*) has only five. Conservation efforts to save the remaining honeycreepers are ongoing. For more about this, see the Appletini'iwi (pg. 128).

This drink is a variation of the Blue Hawaiian, a classic Tiki cocktail that's a bit like a blue Piña Colada. It should not be confused (but often is) with the similarly named Blue Hawaii. This version is lightened up by switching from coconut cream to coconut water.

Blue Hawaiian Honeycreeper

1 oz white rum
1 oz coconut rum
½ oz blue curaçao
1 ½ oz coconut water
1 oz pineapple juice
½ oz lime juice

Combine all ingredients in a cocktail shaker. Add ice and shake until chilled. Strain into a footed pilsner glass filled with ice.

Ingredient Note: If you can, use a quality coconut rum like Clement Mahina Coco or Plantarey Cut & Dry.

Daiqkiwi

Never mind the chicken and the egg—what came first, the kiwi fruit or the kiwi bird? Was one named after the other? Or is the fact that they share a name a coincidence?

It turns out that the bird came first—by about 600 years. The kiwis (five species in the order Apterygiformes) were named by the indigenous Maori people of New Zealand sometime after they arrived on the island in the 1300s. It is thought that the name is onomatopoeic, imitating the birds' call. The fruit, on the other hand, arrived on the island with its European colonists. It was known in English as "Chinese gooseberry," an unwieldy moniker that wasn't helping its popularity as a New Zealand export. Growers took to calling the fruit kiwi due to its resemblance to the brown, hairy birds.

And what unusual birds they are! The five species of kiwi are all flightless and nocturnal. Their feathers are more like hair. They have the smallest eyes and largest eggs, relative to their mass, of all birds. In a lot of ways, they are more like mammals than birds. Why are they so very odd?

Due to its isolation, New Zealand has no native ground-dwelling mammals. As a result, birds and reptiles have evolved to fill the ecological niches that were left vacant.

An **ecological niche** is the role a species plays in an ecosystem—where it lives, what it eats, and what eats it.

With no mammals around to eat them, there were plenty of tasty invertebrates available in the soils and leaf litter of ancient New Zealand for an enterprising little bird to find. And once that ancestor of the kiwis began to specialize in nocturnal ground foraging, it no longer needed to maintain keen eyesight or flight. Instead, it came to rely on its sense of smell. Today, kiwis have the second-largest olfactory bulb relative to their mass of all birds.

Here the classic Daiquiri—a simple but delicious mix of rum, lime juice, and sugar—finds its own niche with the addition of fresh kiwi.

Daiqkiwi

2 oz white rum
¾ oz lime juice
¾ oz simple syrup
Half a kiwi (the fruit, not the bird), peeled and sliced

Place kiwi in the bottom of a shaker and muddle well to release its juices. Add remaining ingredients. Fill the shaker with ice and shake until chilled. Fine-strain into a chilled coupe glass. Garnish with a slice of kiwi.

Fun Bird Fact: Can you guess what bird has the largest olfactory bulb? It might surprise you—the turkey vulture (*Cathartes aura*).

Doctor Bird

The spectacular streamertail hummingbird (genus *Trochilus*), also known as the doctor bird, is the national bird of Jamaica. It is **endemic** to the island, meaning it occurs nowhere else. It also holds a very special place in my heart as the first bird species I ever did research on as an undergraduate student of ornithology.

There are actually two species of streamertail: the red-billed streamertail (*T. polytmus*) and the black-billed streamertail (*T. scitulus*). The red-billed birds occur across almost the entire island of Jamaica, save a small area on the far eastern side of the island where black-billed birds are found instead. Where the two forms meet, there is a small area where they interbreed and we find hybrid hummingbirds with bills that are half-black and half-red. We call this a **hybrid zone.**

Hybrid zones are considered "natural laboratories" where biologists can study how species form and change. Hybrid organisms literally blur the line between species and help us better understand what makes them fundamentally different. In the streamertails, it seems that having a bill with an intermediate color makes it hard for hybrid birds to find a mate. So even though the two hummingbirds are genetically almost identical, this one trait keeps them from merging into a single species.[7]

This cocktail is also a hybrid! It's a combination of two classic drinks: the pre-Prohibition Doctor cocktail (rum, lime, and Swedish punsch) and the bitter, tropical Jungle Bird (pg. 45).

Doctor Bird

1 ½ oz pineapple rum
1 oz Swedish punsch
½ oz Aperol
¾ oz lime juice
¼ oz simple syrup

Combine all ingredients in a shaker with ice and shake until chilled. Strain into a chilled coupe glass. Garnish with a wedge of pineapple.

Ingredient Note: Swedish punsch is a bottled version of the spiced rum-based punches that were popular in the eighteenth and nineteenth centuries. It can be consumed like a cocktail on its own, but it's more often used as an ingredient. Kronan is the easiest brand to find in the United States. For pineapple rum, it's hard to beat Plantarey's Stiggins Fancy Pineapple Rum.

Doradito Mojito

Everyone is familiar with the beautiful songs and calls that birds produce to defend their territory, find a mate, express alarm, and more. These are produced using their syrinx, an organ at the base of their trachea. But some birds also use *other* parts of their bodies to add sounds to their calls or displays. Sounds that birds use for communication that are not produced by the syrinx are known as **sonations.**

The crested doradito (*Pseudocolopteryx sclateri*), a tyrant flycatcher native to South America, is one such bird. Male doraditos have five modified primary feathers that they use to create unique sounds in a special flight display.[8] They also snap their bills while singing to add clicking noises to their song.[9]

Doraditos are far from the only birds to produce sonations. Of the birds depicted in this book alone, male red-billed streamertails (Doctor Bird, pg. 81) produce a whirring sound using their primary feathers, barn owls (Free-for-Owl, pg. 110) will click their beaks and tongues, mourning doves (Dove Potion No. 9, pg. 107) make sounds with their wings when they take off, Andean negritos (Negrito Negroni, pg. 197) make wing-whirring sounds, and sage grouse (pg. 170) make a variety of unusual sounds including a *pop* from their large vocal sacs and a *shush* sound made by rubbing their vocal sacs against their wings.

A mojito is a Cuban classic made with rum, lime, and fresh mint. Here, pineapple juice adds a tropical note, while Suze—a gentian liqueur—adds a hint of bitterness and complexity, along with plenty of vivid yellow in the doradito's honor.

Doradito Mojito

2 oz white rum
¼ oz Suze
¾ oz lime juice
½ oz simple syrup
½ oz pineapple juice
8 mint leaves
3 oz club soda

Combine mint leaves and simple syrup in the bottom of a shaker and muddle. Add rum, lime juice, pineapple juice, and Suze. Shake with ice until chilled. Fine-strain into a highball glass filled with ice. Top with club soda and garnish with a mint bouquet.

Ingredient Note: Suze is a bitter, herbaceous gentian liqueur from France. Alternative options are Salers Aperitif and Avéze.

Hot Tody

If you've never been birding in the Caribbean, you might not be familiar with the todies (family *Todidae*), a group of five tiny, long-billed bird species with sparkling emerald plumage native to Cuba, Puerto Rico, Jamaica, and Hispaniola. Each island has its own species of tody except Hispaniola, which has two: the narrow-billed tody (*Todus augustirostris*) and the broad-billed tody (*T. subulatus*). It's unusual for two closely related species to overlap in an area, because they usually compete for resources. The two Hispaniolan tody species have avoided this by specializing on different habitats, a phenomenon called **resource partitioning.** The narrow-billed tody sticks to wet forests at high elevations, while the broad-billed tody can be found in dry habitats at lower elevations. In the places where they do meet, they stay out of each other's way by occupying different parts of the forest. Narrow-billed todies forage lower and in denser foliage than the broad-billed todies, which prefer the thinner vegetation of the overstory.[10]

This Hot Tody is actually a Hot *Toddy*, a cocktail that is at least 300 years old. A mixture of spirit, lemon, sweetener, and hot water, the Hot Toddy is often considered medicinal. This version takes its inspiration from the Puerto Rican tody (*Todus mexicanus*) and a Puerto Rican shot called the Chichaito (equal parts white rum and anise liqueur).

Hot Tody

1 ½ oz aged rum

¾ oz lemon juice

¾ oz honey syrup (pg. 37)

1 dash absinthe

3 oz boiling water

In an Irish coffee mug, combine rum, lemon juice, honey syrup, and absinthe. Add boiling water. Garnish with a slice of lemon.

Macawsmopolitan

One of the most spectacular birdwatching experiences I ever had was on the banks of the Manu River in southern Peru. We gathered in a blind in the dim light of early morning, our eyes fixed on the steep clay riverbank across from us. As the sun slowly rose, dozens of macaws and other parrots gradually descended, gathering on the bank in huge numbers to eat the clay.

For years, it was thought that parrots ate clay (a behavior known as **geophagy**) to help absorb and neutralize toxins in the foods they eat. However, it turns out this is not the case. Research has shown that they are seeking supplemental nutrients, especially salt. The clay licks parrots choose to use contain about four times more sodium than average. They visit these sites more often in the breeding season, when their nutritional needs are high.[11–13] While there are probably multiple reasons for clay consumption, this recent research has changed the way we view this unusual behavior.

We may not need the extra sodium as parrots do, but salt can still do amazing things for a cocktail beyond its usual place on the rim of a margarita glass. It brightens and enhances flavors, balances bitterness, and blends ingredients together. Try it in this riff on Carrie Bradshaw's favorite, the Cosmopolitan. White rum and passion fruit evoke the Peruvian lowlands, and a pinch of salt is the perfect finishing touch.

Macawsmopolitan

1 ½ oz white rum
¾ oz triple sec
¾ oz lime juice
½ oz cranberry juice
½ oz passion fruit juice
1 pinch of salt

Combine all ingredients in a shaker with ice and shake until chilled. Strain into a chilled cocktail glass. Garnish with a wedge of lime.

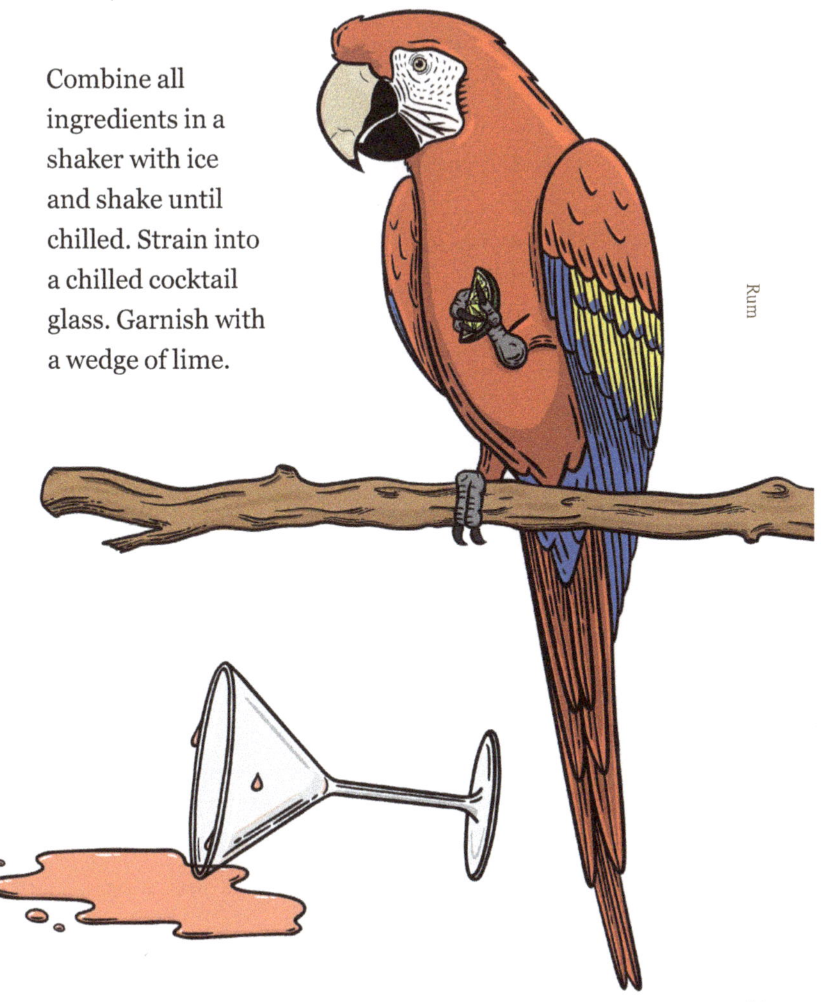

Piña Koel-ada

Being a parent is hard. This is a fact to which I can personally attest, and research in birds supports it. Caring for offspring takes a lot of energy and resources, and wild birds don't have much to spare. Wouldn't it be great to just let someone else do all the work?

That's exactly what the Asian koel (*Eudynamys scolopaceus*) and other members of the cuckoo family do. Cuckoos like the koel are **brood parasites** that lay their eggs in the nests of other bird species and trick those birds into raising their young. As it turns out, most birds aren't too discerning about what their eggs or chicks look like. A roundish object in their nest or a gaping beak is all that is needed to trigger their parental instincts. They will incubate the egg, feed the baby koel—even if it is several times their size!—and raise it to adulthood.

Of course, this is less than ideal for the host parent, who wastes time and energy raising a chick it isn't related to. Check out the Honeyguide cocktail (pg. 60) for more on how these host species have evolved to protect themselves from brood parasitism.

This Piña Koel-ada spices up the classic frozen drink with some flavors from Thailand, right in the middle of the Asian koel's range. Use an Asian rum if you have it; Don Papa from the Philippines is one of the easiest to find.

Piña Koel-ada

2 oz aged rum
2 ½ oz pineapple juice
1 oz cream of coconut
2 tbsp loosely packed fresh cilantro
½–1 Thai chili, seeded and chopped
1 cup ice

Combine all ingredients in a blender and blend until smooth. Pour into a hurricane glass and garnish with a slice of pineapple, a Thai chili, and a sprig of cilantro.

Sazeracket-Tail

The white-booted racket-tail (*Ocreatus underwoodii*) is an extraordinary hummingbird found in the northern Andes. It is aptly named for the fluffy white down around its feet and its distinctive racket-shaped tail feathers. These racket plumes are not unique to hummingbirds; they're found in a number of other birds, including the racket-tailed parrots, some motmots, some paradise kingfishers, and the greater racket-tailed drongo. They've existed for a long time. Fossils and amber deposits from the Cretaceous period (~99 Ma) show that early birds like *Confuciusornis* and *Junornis* had similar tail feathers.[14]

In most birds with racket plumes, the feathers form that way naturally. Though the shaft looks bare, there are actually microscopic feather barbs along its length, meaning that racket plumes aren't as different from other feathers as they appear.[15] The exception is the motmots. They grow ordinary tail feathers with weakened barbs along their length that are prone to rubbing and falling off, leaving them with racket plumes.[16] Just two different ways to get to the same flashy tail feathers!

The Sazerac, a classic New Orleans cocktail, is also traditionally made two ways: with rye whiskey or with Cognac. However, it's even more delicious when made with rum. Try Diplomatico Reserva from Venezuela, which is made right at the foot of the mountains where you'll find the white-booted racket-tail.

Sazeracket-Tail

2 oz aged rum
¼ oz passion fruit syrup (pg. 38)
1 dash Angostura bitters
1 dash orange bitters
Absinthe rinse

Spritz or rinse a small rocks glass with absinthe. In a mixing glass, combine rum, passion fruit syrup, and bitters. Add ice and stir until chilled. Strain into the prepared glass and garnish with an orange twist.

Toucan Play at That Game

The toco toucan (*Ramphastos toco*) is one of the most recognizable birds in the world. As the largest of all the toucans, it has the highest beak-to-body ratio of all birds. But why does it have that big, beautiful beak? One answer may surprise you. It doesn't have to do with the type of food the toucan eats or attracting mates. It's all about temperature.

Toucan bills have a network of blood vessels running under their surface. Scientists found that the beak of the toco toucan is significantly cooler than its body temperature, and that blood flow to the beak increases at warmer temperatures and decreases in the cold.[17] As blood travels through the beak, it cools down before reentering the body, acting as a cooling mechanism. The toucan's beak is essentially a radiator.

Toucans aren't the only animals with this type of adaptation. It's well-documented that animal appendages like ears, horns, and even limbs act as sources of heat loss. There is an ecogeographical "rule" associated with this concept called **Allen's Rule,** which states that animals in colder climates tend to have shorter appendages, and animals in warmer climates have longer ones (see Bobwhite Russian, pg. 131, to learn about another one).

Thermoregulation is important for cocktails, too. One reason bartenders use chilled glassware is to help keep your drink colder, longer. Similarly, hot water is often poured into glass mugs before serving an Irish coffee or other warm cocktail. But did you know that there is a cocktail that is not served hot or cold? A scaffas is a type of drink that is meant to be served at room temperature. A bit of water is usually added to achieve the proper amount of dilution.

Toucan Play at That Game

1 oz aged rum (preferably Jamaican)
½ oz pineapple rum
½ oz amontillado sherry
¼ oz orange curaçao
¼ oz Amaro Nonino
½ oz water
1 dash Angostura bitters
1 dash orange bitters

Add all ingredients to a small rocks glass and stir briefly. Garnish with an orange twist, discarded.

Vangastura Sour

The vangas (family Vangidae) are an unusual group of birds that provide a spectacular example of an **adaptive radiation.** These are two words that get evolutionary biologists very excited. An adaptive radiation is when several populations become different from one another by evolving to specialize on different niches or environments very quickly. The classic example is Darwin's finches, which evolved different bill sizes and shapes to specialize on different food sources.

The fact that such a famous radiation took place in the Galapagos Islands is no accident. Adaptive radiations often occur on island chains, where geographic isolation stops these different groups from interbreeding, speeding evolution along. The vanga radiation is even more impressive because it occurred on a single (admittedly large) island: Madagascar. Isolated there, the ancestors of the vangas diversified greatly, evolving into forty species you would never guess were so closely related, from the nectar-eating sickle-billed vanga with its thin, arched bill to the insectivorous helmet vanga (*Euryceros prevostii*, pictured), which sports a brilliant blue bill with a bony protrusion from its head called a casque.

Such an unusual group of birds calls for an unusual cocktail. Angostura bitters are a cocktail ingredient normally added to drinks in one or two dashes, but bartender Giuseppe Gonzales's Trinidad Sour turned

this concept on its head in 2009. It contains a full ounce of Angostura bitters, treating it as a base for the drink. Even cocktail ingredients can adapt!

Vangastura Sour

- ¾ oz aged rum
- ¾ oz Angostura bitters
- ¾ oz cream of coconut
- ¾ oz lime juice

Combine all ingredients in a shaker with ice. Shake well. Fine-strain into a chilled Nick and Nora glass. No garnish.

Fun Cocktail Fact: The Trinidad Sour is actually a variation of the Trinidad Especial, a cocktail by Italian bartender Valentino Bolognese that won the European Finals of the Angostura Global Cocktail Challenge in 2008. However, it was Gonzales's Trinidad Sour that became an instant classic the following year.

Tequila

Boom Chachalaca

All baby birds are not created equal. A baby robin hatches blind and featherless, unable to fly, feed itself, or flee from predators. It is entirely dependent on its parents. This is true for most baby birds, and it may also bring to mind another very familiar species: humans. Both robins and humans are **altricial** species, meaning they are born at an earlier stage in their development, requiring significant parental care to survive and grow.

In contrast to altricial birds, **precocial** birds like ducklings or baby chicks are born ready to fend for themselves. They may still need some help and instruction from their parents, but they can leave the nest, forage for their own food, and run and hide from danger within days or even hours of hatching. Waterfowl, landfowl, shorebirds, cranes, and ratites (ostriches and their relatives) are some of the groups that have precocial young.

The chachalacas (genus *Ortalis*) are large forest birds found in Central and South America, Mexico, and southern Texas. They are related to other landfowl like chickens, pheasants, turkeys, and quails. Like these others, they have precocial young that hatch highly developed. Baby chachalacas can cling to tree branches as soon as they are hatched (an important skill for a precocial bird that nests in trees!) and can fly within a few days.

Chachalacas are found throughout Mexico, so this cocktail takes a little inspiration from one of the country's most famous drinks, the Paloma (pg. 46). The smoky flavor of mezcal is reminiscent of the aftereffects of an explosion. Boom!

Fun Bird Fact: There is a group of birds called megapodes that are **super precocial.** They bury their eggs under decaying vegetation, which keeps the eggs warm. The parents do not incubate the eggs, and the chicks are entirely independent as soon as they hatch. Some even hatch with their full set of flight feathers!

Boom Chachalaca

1 ½ oz mezcal
½ oz Aperol
2 oz grapefruit juice
½ oz lime juice
¼ oz cinnamon syrup
Club soda, to top (about 2 ½ oz)

Combine mezcal, Aperol, grapefruit juice, lime juice, and cinnamon syrup in a shaker. Add ice and shake until chilled. Strain into a highballglass and top with club soda. Garnish with a grapefruit slice and a bouquet of fresh mint.

CINNAMON SYRUP

1 cup sugar
½ cup water
3 cinnamon sticks, broken into pieces

Combine sugar and water in a saucepan over medium heat. Bring to a simmer, stirring until the sugar is dissolved. Add cinnamon sticks and stir. Let the mixture simmer for a couple of minutes, then cover and remove from heat. Let cool completely. Strain out the cinnamon sticks. Store in the refrigerator for up to two weeks. Try it in an Old-Fashioned, Boukman Daiquiri, or Jet Pilot.

Dove Potion No. 9

Coming up next, it's *CSI: Forensic Ornithology.*

In 1960, Eastern Airlines flight 375 crashed into Boston Harbor shortly after taking off from Logan Airport. As officials investigated the cause of the crash, they found dead birds on the runway and feathers in the plane's machinery. They sent samples to the Smithsonian National Museum of Natural History for identification. Researcher Roxie Laybourne quickly identified them as European starlings. The plane had hit a massive flock of them shortly after takeoff.

Laybourne had been interested in the microscopic structure of feathers, particularly downy feathers, for many years. She could look at a single feather under a microscope and identify what species it came from. But it wasn't until that fateful plane crash that she—and the US government—realized how important her work could be. Within a few years, Laybourne was running a lab at the Smithsonian with the sole task of identifying bird strike remains (playfully dubbed "snarge" by military pilots). By identifying the species of the culprits, airports can take preventative measures to keep those birds out of the area and prevent future accidents.

Laybourne died in 2003 at the age of ninety-two. Today, her former protégé Carla Dove runs the lab, which now uses DNA barcoding in addition to microscopic feather anatomy. They analyze over 10,000 bird strike samples a year for the Federal Aviation Administration. The species most often involved? The humble mourning dove (*Zenaida macroura*).[18] Fortunately, these incidents are not usually catastrophic due to the mourning dove's small size. It's larger waterfowl that tend to cause major crashes, such as the Canada geese that forced US Airways Flight 1549 to land in the Hudson River.

What better drink for the number one cause of bird-related air accidents than the Aviation? The original version of this cocktail (gin, lemon, maraschino liqueur, and crème de violette) was named for its sky-blue hue. What this tequila-based recipe lacks in color, it makes up for in flavor.

Dove Potion No. 9

1 ½ oz blanco tequila
¾ oz maraschino liqueur
¼ oz Ancho Reyes
½ oz lemon juice

Combine all ingredients in a cocktail shaker with ice. Shake until chilled. Strain into a chilled cocktail glass or coupe and garnish with a brandied cherry.

Ingredient Note: Ancho Reyes is a surprisingly versatile chile-flavored liqueur. Try adding it to a Paloma or Margarita.

Free-for-Owl

It's not easy hunting in the dark, but owls like the American barn owl (*Tyto furcata*) make it look like a breeze. Their soft feathers allow them to fly near-soundlessly. Their facial feathers are arranged into a disc that helps funnel sound into their ears. Their hearing is so acute and directional that they can locate prey beneath vegetation or snow and snatch it up with their long legs and talons. But it's their eyes that give them the most advantages in the dark.

Owl vision is specially adapted for low light. Their eyes are tube-shaped rather than round, helping to funnel light to their retinas. Because a tube-shaped eye can't swivel around in a socket, they are held in place by bony rings and can only look forward. To accommodate for this, owls can rotate their heads 270 degrees, giving them a wide range of both vision and hearing.

Once that light is funneled to the retina, it hits the owl's photoreceptors. These cells come in two types: rods and cones. Cones provide color vision (see Rosélla, pg. 203), while rods are more light-sensitive. Birds that are active during the day have many more cones than rods. The retina of an owl, on the other hand, is packed with

rod cells. While this means they essentially see in black and white, it gives them excellent night vision.

I hope your night vision is good, because you won't be sleeping after this cocktail. The vodka in this Espresso Martini is swapped for a far more flavorful reposado tequila, and a dash of bitter Mexican Fernet Vallet rounds out the flavors.

Free-for-Owl

1 ½ oz reposado tequila
¾ oz coffee liqueur
¼ oz Fernet Vallet
1 oz espresso, cooled
½ oz simple syrup
1 dash vanilla extract

Combine all ingredients in a shaker with ice and shake until chilled. Strain into a cocktail glass and garnish with star anise.

Ingredient Note: Fernet Vallet is a bitter Mexican liqueur. If you don't have it, try replacing it with another amaro or omitting it from the recipe. Fernet Branca is an option if you're a fan, but its minty notes may feel a bit out of place in the drink.

Fulmargarita

The fulmars (genus *Fulmarus*) are seabirds in the order Procellariiformes. Birds in this group, which also includes petrels, shearwaters, storm petrels, and albatrosses, are known as **"tube-nosed" seabirds** due to their unique, tube-shaped nasal passages. These special noses do a lot. The fulmars' nasal glands remove salt from their bodies and excrete it through their nostrils as a saline solution, allowing them to drink seawater. The tube shape gives them an excellent sense of smell, helping them locate their prey (fish, squid, and crustaceans) at sea.

As all that seafood is digested, Procellariiformes like the fulmars produce a stomach oil in their foregut. It has an extremely high caloric density and helps to sustain the birds on long flights. It is also fed to chicks. But some species, including the fulmars, use it for something else: defense. They can forcefully regurgitate this foul-smelling stomach oil to deter predators (eww).

Luckily, there's nothing foul about this marine-inspired margarita. It's the color of the Arctic sea, with a briny salt rim. Kelp powder gives it a hint of umami flavor that blends in unexpectedly well.

Fulmargarita

2 oz blanco tequila
¾ oz blue curaçao
1 oz lime juice
¼ oz agave nectar
1 pinch kelp powder

Run a wedge of lime along the edge of a chilled coupe glass and dip it in sea salt to coat. In a shaker, combine tequila, lime juice, curaçao, agave, and kelp powder. Fill the shaker with ice and shake until chilled. Strain into the prepared glass.

Mad Cowbird

I've talked a lot about brood parasites in this book (see the Honeyguide, pg. 60, and the Piña Koel-ada, pg. 92). As you hopefully recall, brood parasites are birds that lay their eggs in the nests of other species. Most hosts will raise and care for the brood parasite's young, at great cost to themselves. In most cases, ornithologists believe that this is because they can't tell the difference. They're being tricked. But sometimes, it's because the brood parasite makes an offer the host bird can't refuse—a theory biologists have dubbed the **mafia hypothesis.**

The brood parasitic brown-headed cowbird (*Molothrus ater*) doesn't just drop off her eggs and run. She will return to the nests where she deposited their eggs to make sure those eggs are still there. If she finds that the host bird has rejected the egg and tossed it out of the nest, she will retaliate by destroying the nest and all of the host's eggs. Raising a baby cowbird may take a lot of energy and resources, but losing everything and starting over from scratch is far worse. So the host species accept the cowbird eggs and raise the baby cowbirds as their own. This same retaliatory behavior has also been observed in cuckoos.

What better inspiration for this recipe than the Godfather cocktail? Made of equal parts Scotch and amaretto, it's too sickly sweet for most modern palates.

This slow sipper is more balanced, and tequila is a fantastic partner for amaretto. Try it...or else!

Mad Cowbird

- 2 oz reposado tequila
- ¼ oz amaretto
- 1 barspoon allspice dram
- 2 dashes Angostura bitters
- 2 dashes orange bitters

Combine all ingredients in a mixing glass filled with ice. Stir until chilled. Strain into a rocks glass over one large ice cube. Garnish with an orange twist.

Mai Tyrant

Do birds learn their songs, or are they born knowing how to sing? The answer is: it depends on the bird.

The best singers among birds are the **oscine** passerines. This is the group we call songbirds. They have a vocal organ called the syrinx that enables them to sing complex, melodious songs. They learn these songs from their parents and neighbors when they're young. If they're raised in isolation, they won't learn to sing properly.

There's also a group known as the **suboscine** passerines. These are also perching birds, but they have a simpler syrinx that can't quite produce the variety of sounds that an oscine's can. They *don't* learn their songs—they are born knowing them. Their songs are essentially encoded in their DNA! They are mostly found in South America and Asia, but one group makes it up to the United States as well: the tyrant flycatchers (family Tyrannidae), like the beautiful vermilion flycatcher (*Pyrocephalus obscurus*). Something to think about the next time you hear one sing!

The Mai Tai is a quintessential Tiki cocktail that is traditionally made with rum, orange curaçao, lime juice, and orgeat (an almond syrup). Bars tend to take liberties with this recipe, making unnecessary additions like pineapple juice, orange juice, or grenadine. If you're served a bright red Mai Tai in the wild, it was not made properly.

But if you're served *this* bright red Mai Tai, expect a delicious drink inspired by the vermilion flycatcher's plumage and its distribution all over Mexico.

Mai Tyrant

1 oz reposado tequila
1 oz mezcal
½ oz orange curaçao
½ oz Aperol
1 oz lime juice
½ oz orgeat

Combine all ingredients in a shaker. Add ice and shake until chilled. Strain into a rocks glass filled with crushed ice. Garnish with a bouquet of fresh mint.

Fun Bird Fact: Of course, there's also the non-passerines or non-perching birds like waterfowl, birds of prey, hummingbirds, parrots, and woodpeckers. Some of them learn their vocalizations and some don't—there's a lot of variation!

Porn Starling Martini

If you pay any attention to birds at all, you've certainly seen a European starling (*Sturnus vulgaris*). More likely, you've seen thousands. But as you might guess from their name, they don't belong in the United States at all. How did a European species become one of the most numerous and successful birds in North America?

A species that establishes itself in a new place after being brought there by humans is known as an **introduced species.** We know now that introduced species can be a huge issue for the local environment, outcompeting native species for food, nesting sites, and other resources. But this was poorly understood in the mid-1800s, when Americans sought to introduce foreign birds to the country for their beauty, familiarity, and/or economic benefit. Acclimatization societies were formed with the sole purpose of establishing non-native species in the country. European starlings did not colonize North America by escape or accident; they were deliberately released. And now there are millions of them.

As the nineteenth century came to a close, the negative effects of overhunting, introduced species, and habitat destruction began to become apparent. Once-abundant native species like the passenger pigeon and Carolina parakeet were well on their way to extinction.

PREMIUM Miller BREWED
HIGH LIFE
EST° 1903
The Champagne of Beers
7 FLUID OUNCES

The Lacey Act of 1900 prohibited the importation and introduction of foreign species into the United States without a permit, but as far as starlings were concerned, the damage was done.

This cocktail has a somewhat inappropriate name (perhaps fitting for a bird with the species name *vulgaris*). The Porn Star Martini was created by bartender Douglas Ankrah in 2002. The passion-fruit-laced vodka sour is normally served with a sidecar of champagne; however, for the common-as-dirt European starling, this tequila version is accompanied by a pony of Miller High Life (or feel free to substitute a so-bad-it's-great cheap beer of your choice).

Porn Starling Martini

1 ½ oz reposado tequila
½ oz passion fruit liqueur
1 oz lime juice
¾ oz falernum syrup
Miller High Life pony

Combine tequila, lime juice, falernum syrup, and passion fruit liqueur in a shaker. Add ice and shake until chilled. Strain into a chilled coupe and serve with the Miller High Life as a sidecar.

Ingredient Note: Falernum syrup is a Caribbean syrup made with sugar, lime, spices, ginger, and/or almond. Alcoholic falernum liqueurs are also sold and could be used as a substitute.

Tequila Sunbird

Birds have adapted to make use of just about every food source available on the planet, from carrion to cactus and blood to beeswax. So, it should come as no surprise that a number of bird families have come to make nectar a substantial part of their diet. In the New World, of course, we have the hummingbirds; in the Old World, their counterparts are the sunbirds (family Nectariniidae).

Like hummingbirds, sunbirds are small, beautifully iridescent, and **nectarivorous** (meaning they eat nectar from flowers). Nectar is an abundant food source, particularly in the tropics, and contains high amounts of simple sugars. However, a nectar-rich diet has its challenges, and sunbirds like the fire-tailed sunbird (*Aethopyga ignicauda*) pictured here have evolved a number of special adaptations.

Nectar is a fairly dilute liquid, so birds need to drink a lot of it to meet their caloric requirements. But liquid is heavy, so they have to process it quickly and dump it in order to fly. For this reason, nectarivorous birds have short digestive systems that break down and absorb sugars extremely efficiently, and kidneys that conserve electrolytes while processing high volumes of fluid. Despite all these adaptations, they still can't survive on nectar alone. Even hummingbirds must supplement

their diets with insects in order to obtain vital nutrients such as nitrogen, protein, amino acids, and electrolytes. Some members of the sunbird family do this so much that they are known as spiderhunters.

To extract a meal of nectar from a flower, a sunbird pokes its beak in and reaches out with its brush-tipped tongue. The edges of the tongue are curled inward, creating a tube that pulls the nectar up through capillary action, as the birds cannot create suction themselves. Imagine being able to dip the tip of your tongue into your cocktail and have it flow right up into your mouth!

Sadly, you'll need a straw for this riff on the classic Tequila Sunrise. The usual combination of tequila, orange juice, and grenadine has always felt a bit lackluster, but this simple variation adds some fizz and floral flavors (for the sunbirds!) without eliminating the beautiful sunrise effect of the original.

Tequila Sunbird

1 ½ oz blanco tequila
1 ½ oz chilled hibiscus tea
4 oz Italian orange soda

Fill a highball glass with ice. Add tequila and soda, then carefully top with tea. Stir before you enjoy.

Vodka

Appletini'iwi

Elsewhere, I have discussed the plight of the Hawaiian honeycreepers, a family of birds endemic to Hawaii that is rapidly going extinct due to deforestation and avian malaria (see Blue Hawaiian Honeycreeper, pg. 75). For one Hawaiian honeycreeper, the I'iwi (*Drepanis coccinea*), hope is coming from a very unexpected place.

Like its brethren, the I'iwi has suffered a major population decline in the last few hundred years. Its once-expansive range has shrunk to patches of wet, high-elevation forest where the mosquitoes carrying avian malaria do not occur. In order for the I'iwi to return to its lowland home, it needs to find a way to fight this disease. When we humans find ourselves in a similar situation, we develop vaccines. But I'iwi are not likely to stop in at their local CVS for a yearly shot. So scientists have come up with another potential solution: editing the genome of the I'iwi to be resistant to avian malaria.[19]

While this was science fiction just a decade or two ago, genome editing is rapidly becoming a viable possibility in wild animals with tools like CRISPR. Attempting to accelerate evolution in this way is known as **facilitated adaptation.** At this point it remains largely theoretical, though introducing beneficial traits into endangered populations has been done by bringing individuals that already have them into the population. No one knows what advances the next two decades will bring—and whether the I'iwi will still be around to see them.

Speaking of adaptation, the Appletini is a cocktail that certainly needs to adapt to the times. I'iwis may love sugar, but too much of it can ruin a drink, and the Appletini is often made with overly sweet apple liqueurs full of artificial colors and flavors. Given the I'iwi's Hawaiian home, the "apple" in this Appletini'iwi is pineapple instead.

Appletini'iwi

2 oz vodka (try PAU Hawaiian vodka)
¾ oz lemon juice
½ oz pineapple juice
¼ oz hibiscus syrup

Combine all ingredients in a shaker. Add ice and shake until chilled. Strain into a chilled cocktail glass and garnish with a pineapple wedge.

HIBISCUS SYRUP

1 cup brewed hibiscus tea (sometimes called sorrel tea)
1 cup sugar

In a saucepan, combine tea and sugar. Bring to a simmer, stirring frequently, until sugar is dissolved. Let cool completely before using. Store in the refrigerator. Try it in a Margarita or Daiquiri.

Bobwhite Russian

The northern bobwhite (*Colinus virginianus*) is a species of quail found from southern Canada down to Central America. Like many bird species with large ranges, it experiences a wide variety of climates. Some populations live in areas that experience harsh, cold winters and mild summers, while others dwell perpetually in the tropical heat. If you compare the average weight of bobwhites in these two places, you'll find that the northern birds are much heavier than those in the south.

In 1847, a German scientist named Carl Bergmann published a paper showing that this trend is actually widespread: animals in colder places generally tend to be larger than their counterparts in warmer places. This has come to be known as **Bergmann's Rule.** Bergmann suggested that larger animals have a lower surface-to-volume ratio to lose less heat, while smaller animals have a higher ratio to help them cool off. A similar principle applies to cocktail ice: small cubes with higher surface-to-volume ratios melt faster, while large spheres or cubes take longer.

You'll probably drink this delicious Bobwhite Russian so quickly that your ice won't have much time to melt. Creamy, cinnamon-laced horchata liqueur (a nod to the Mexican portion of the northern bobwhite's range) is right at home among the typical ingredients of a classic White Russian.

Fun Science Fact: Carl Bergmann was a busy guy; he also identified rods and cones as the photosensitive cells of the retina (see Free-for-Owl, pg. 110) and coined the term "fovea" (see Hawkward Position, pg. 222).

Bobwhite Russian

1 oz vodka
1 oz horchata liqueur
½ oz coffee liqueur
1 oz cream

Combine all ingredients in a shaker with ice and shake until chilled and frothy. Strain into a rocks glass filled with ice and garnish with freshly grated nutmeg.

Brambling

Estimating bird counts is a difficult skill to master. If you're a birder who likes to keep track, you probably have tricks to help you figure it out, such as breaking the flock into units of ten or 100 and counting those, or counting all the birds in a fraction of the flock and multiplying. With practice, you can use these techniques to count huge flocks with thousands or even tens of thousands of birds. But what about hundreds of thousands? What about...millions?

It's a relatively rare event when that many birds gather in one place, and certain species are more prone to do it than others. One bird that certainly doesn't mind a crowd is a lovely little Eurasian finch called the brambling (*Fringilla montifringilla*). Bramblings eat beech nuts in the winter, and during "mast years" when the beeches in a certain area produce large amounts of fruit, the birds will congregate in huge roosts to take advantage of it. Some recent examples of this occurred in Sweden in 2019–2020 and in Slovenia in 2004–2005. The largest gathering of birds ever recorded was a flock of *70 million* Bramblings in Switzerland. Since that report was made in the winter of 1951–1952 and (as we've just established) estimating bird numbers is difficult, it's possible there weren't quite that many. But to even guess 70 million means there were a *lot* of birds there.

Birds that gather in these huge flocks aren't just taking advantage of abundant food. The large numbers provide additional protection from predators, help them keep warm, and possibly even give them a chance to exchange information. Whatever the reason, it certainly makes for a breathtaking spectacle!

The Bramble is a "new classic" cocktail created by bartender Dick Bradsell in the 1980s. (He also shook up the first Espresso Martini.) Here the gin in the original recipe is swapped for vodka in honor of the bramblings' chosen winter roosts in Sweden and Eastern Europe, and ginger syrup adds a bit of zip.

Brambling

2 oz vodka
½ oz crème de mûre
1 oz lemon juice
½ oz ginger syrup (pg. 37)

Combine vodka, lemon juice, and ginger syrup in a shaker with ice. Shake until chilled. Strain into a rocks glass filled with crushed ice. Gently top with crème de mûre. Garnish with a lemon wheel and a blackberry.

Coconuthatch

North American birders are very familiar with the nuthatches (genus *Sitta*), small birds that cling to the trunks of trees as they forage for insects. Nuthatches are actually found across the Northern Hemisphere and even in parts of India and southeast Asia. Most of the twenty-nine species in this group look fairly similar to those we're used to seeing: a grey back, white belly, and a black or brown cap. *Maybe* they'll have an eye stripe or some reddish color on the belly. But the closer you get to the equator, the more colorful and unusual the nuthatches start to look, until you find spectacular species like the velvet-fronted nuthatch, the sulphur-billed nuthatch, and the blue nuthatch (*Sitta azurea,* pictured). If you think about it, this is true of birds in general—they seem to be more colorful in the tropics. Why?

The truth is, we don't know for sure. There are many theories. One is that there are more resources available in the tropics, and colors are energy-intensive to create. More specifically, tropical foods are rich in carotenoids, an important component of some bird colors (see Scarlet Rye-bis, pg. 172, and Booby Trap, pg. 56). Another is that males need to be brighter to catch the eye of females in the thick vegetation. Maybe the most interesting theory is that birds aren't *actually* more colorful in the tropics at all. For every bright parrot and tanager, there are also a number of plain brown and grey birds that just don't get as much attention.

Because of all the resources available in the tropics, there are more species there *in general*, both colorful and not. This is even true in the nuthatches—there are some plainer species that overlap with the flashy ones.

Cocktails definitely get flashier and more colorful as they become tropical—just go to any Tiki bar and you'll see. But, as in reality, some plainer examples fly under the radar. This subtly tropical Martini is one such drink.

Coconuthatch

- 2 ½ oz coconut-washed vodka
- ¾ oz lemongrass-infused dry vermouth
- 1 dash yuzu bitters

Combine all ingredients in a mixing glass filled with ice. Stir until well-chilled. Strain into a chilled cocktail glass or coupe and garnish with a lime twist.

COCONUT-WASHED VODKA

- 1 cup vodka
- 2 tbsp coconut oil

Warm the coconut oil until it is completely liquid. Pour 1 cup of vodka into a jar and add the melted coconut oil. Seal the jar and shake briefly. Let sit at room temperature for several hours, shaking occasionally. Then transfer to the freezer overnight. Once the coconut fats are frozen and solid, strain through a coffee filter to remove them.

LEMONGRASS-INFUSED DRY VERMOUTH

1 cup dry vermouth
2 stalks of fresh lemongrass

Add dry vermouth to a jar. Cut 2 stalks of lemongrass into 1-inch pieces. Cut these pieces in half lengthwise. Add them to the vermouth. Seal the jar and shake briefly. Let sit in the refrigerator overnight before straining out the lemongrass. Keep infused vermouth stored in the fridge.

Grain of Saltator

If you've never opened a field guide to birds before, you might be quite puzzled about how it's arranged. Birds that look similar and do similar things are generally placed together, but not always. Hawks and falcons, for example, are separate in some guides. Swifts and swallows are always very far apart. Why are parrots next to doves and hummingbirds next to swifts? And if there *is* logic to this organization, why isn't it consistent across different guides?

Bird field guides are usually arranged in **taxonomic order.** This means that closely related birds are placed together, and the overall order is determined by how early each group of birds branched off from the avian family tree. The ratites and tinamous come first, followed by the waterfowl and land fowl, and so on. The perching birds (Passerines) are always last.

The reason why different versions of field guides may have the birds in a different order is that our understanding of bird relationships is always changing as we find out more about them. DNA sequencing in particular has drastically changed how we view the avian tree of life. The saltators (genus *Saltator*) are just one example of this. Ornithologists always thought they were related to the cardinals, and it wasn't until their DNA was sequenced and compared to that of other birds that we discovered they are actually tanagers.[20] Naturally, no publisher is going to rush to make these corrections,

especially since they happen quite frequently now, but updated relationships may be reflected in the next edition.

With this in mind, you should always take your bird taxonomy with a grain of salt—and sometimes your cocktails, too! A pinch of salt is a particularly nice addition to this refreshing cucumber recipe.

> *Fun Bird Fact:* Another more famous example of this is Darwin's finches, which are not finches at all—DNA sequencing has told us that they are also in the tanager family. Somehow Darwin's tanagers just doesn't have the same ring to it.

Grain of Saltator

1 ½ oz vodka
½ oz elderflower liqueur
1 oz lemon juice
½ oz simple syrup
5 slices cucumber
5–6 basil leaves
1 small pinch of salt

Combine cucumber, basil, salt, and simple syrup in the bottom of a shaker. Muddle well to release the cucumber's juice. Add remaining ingredients. Fill the shaker with ice and shake until chilled. Fine-strain into a rocks glass filled with ice. Garnish with a slice of cucumber and a bouquet of fresh basil.

Hoopoe Coupe-o

There are few birds as unique and charismatic as the Eurasian hoopoe (*Upupa epops*). With its regal crest, long bill, and striking plumage, it's no wonder that the hoopoe features in the mythology and tradition of many cultures within its massive range. It is depicted on the walls of ancient Egyptian tombs and mentioned in both the Quran and the Old Testament of the Bible. In the Persian poem *The Conference of the Birds*, the wise hoopoe leads the other birds on a quest. In Medieval Europe, it was thought that hoopoe blood or a stone from its nest could cause hallucinations or induce someone to tell the truth.

In Greek mythology, the hoopoe is associated with King Tereus, a not-so-great guy who rapes his wife's sister and then cuts out her tongue. His wife, understandably angry, responds by (somewhat less understandably) killing her own son and serving him to Tereus for dinner. Furious, Tereus tries to kill both women, but before he can, the gods transform all three of them into birds. Tereus is turned into a hoopoe. It is in this form—though missing most of his feathers—that he appears in the play *The Birds* by Aristophanes. He helps two Athenian men convince the rest of the birds to build a city in the sky called Cloud Cuckoo Land.

Creamy, tangy, and aromatic, this cocktail features Greek ingredients like pistachio, cardamom, and mastiha, a liqueur made with mastic resin on the island of Chios. If you travel to Greece, you may be offered a small glass of mastiha as a welcome drink or a digestif after dinner. Kleos and Skinos are two brands that you can find in the US.

Hoopoe Coupe-o

1 ½ oz vodka
¾ oz mastiha
1 oz lime juice
½ pistachio-rose orgeat
1 dash cardamom bitters

Combine all ingredients in a shaker with ice and shake until chilled. Strain into a chilled coupe. No garnish.

PISTACHIO-ROSE ORGEAT

(adapted from Amanda Marsteller for Liquor.com)

2 cups raw shelled pistachios
1 ½ cups sugar
1 ½ cups water
1 oz vodka
1 tsp rosewater

Pulse the pistachios in a food processor until they are finely ground. Combine sugar and water in a saucepan over medium heat and bring to a simmer, stirring frequently, until sugar is completely dissolved. Add the ground pistachios. Reduce heat to low and simmer for three minutes, stirring occasionally. Cover and remove from heat. Let sit until completely cool. Strain the liquid through cheesecloth. Add the vodka and rosewater, stirring to combine. Can be stored in the refrigerator for up to two weeks.

Moscow Murre

The murres (genus *Uria*) are two species of Arctic seabirds. Along with their relatives, the auks and puffins, they're essentially the Northern Hemisphere's answer to penguins. While awkward on land, they are excellent swimmers and divers. They even share the penguins' black and white coloration. But unlike their southern counterparts, they can fly. This allows them to nest on high cliffs, safe from potential predators. However, their eggs face another danger, just as deadly: rolling off the cliffside to smash on the rocks below.

In response to this risk, murres have evolved an unusual, conical egg shape that is narrow on one end and broad on the other. The eggs roll in an arc rather than a straight line, so they're less likely to roll off a cliff, and their unique shape increases the amount of surface area touching the ground, providing additional stability.[21] The eggs also exhibit a huge amount of variation in their colors and patterns. This helps individual murres recognize their own eggs in a crowded nesting colony.[22] Even before they hatch, the murres are perfectly adapted to their environment.

The similarity between the word "murre" and the French word *mûre*, meaning "blackberry," is purely coincidental, but too great not to lace this riff on a Moscow Mule with a healthy pour of crème de mûre, a jammy blackberry liqueur.

Moscow Murre

1 ½ oz vodka
¾ oz crème de mûre
¾ oz lime juice
3 oz ginger beer

Combine vodka, crème de mûre, and lime juice in a shaker with ice and shake until chilled. Strain into a Moscow Mule mug filled with ice. Top with ginger beer and stir briefly. Garnish with a bouquet of fresh mint and a blackberry.

Vesper Sparrow

Bond. James Bond. Ornithologist.

Ian Fleming, author of the original *James Bond* novels, was living in Jamaica when he started writing them. He was also an avid birder. While trying to name the main character of his new book series, he recalled the name of the author of one of his field guides, *Birds of the West Indies*. It was exactly what he was looking for: short, simple, and masculine. And so, the British spy James Bond was born—named after an American ornithologist.

The real Bond was born in Philadelphia, where he became the curator of ornithology at the Academy of Natural Sciences. He was an expert on Caribbean birds, and his book was a seminal reference for the birds of the region (Fleming once referred to it as "one of [his] bibles").[23] Bond had no idea that Fleming had appropriated his name until several of the popular books had been published. His wife corresponded with Fleming, who jokingly gave Bond permission to use his own name as he saw fit. "Perhaps one day he will discover some particularly horrible species of bird which he would like to christen in an insulting fashion," Fleming suggested.[23]

In *Casino Royale*, the first *James Bond* book, the spy invents a cocktail he names the Vesper, after his love interest Vesper Lynd. This cocktail is also dedicated to a beautiful creature, the North American vesper sparrow (*Pooecetes gramineus*). It gets its name from the beautiful song it sings in the early evening, when a Christian prayer service called vespers is traditionally held. The Vesper Sparrow is a delightful cocktail to drink during this time of day, and is meant to evoke the American prairies at sunset.

Vesper Sparrow

1 ½ oz vodka
½ oz peated Scotch
¼ oz Lillet blanc
Olive, lemon peel, and fresh sage for garnish

Combine vodka, Scotch, and Lillet in a mixing glass with ice. Stir until chilled. Strain into a chilled cocktail glass. Express a lemon peel over the drink and rub it along the rim of the glass. Garnish with an olive (a little olive juice would not go amiss) and a sage leaf.

Ingredient Note: Use an Islay Scotch, such as Laphroaig, Lagavulin, or Ardbeg.

Whiskey

Apple Eider

Every time I mention a bird species in this book, I include two italicized words in parentheses behind it. If you're a birder or a scientist or even if you remember your high school biology, you probably know what this is: it's the bird's **scientific name.**

Scientific names are important because they are standardized around the world. Take the king eider, for example. It's called eider à tête grise in French, éider real in Spanish, and Prachteiderente in German. You can see how things could quickly get confusing.

Enter the scientific name, an international standard that is at least equally confusing to everyone. This format was created by a Swedish naturalist named Carl Linnaeus in 1735. We still use many of the names he gave to plants and animals today. Each one is made up of two words, the animal's **genus** and its **species.** A genus is a group of closely related species. Both names usually come from Latin or Greek. They are always written in italics, and the genus is capitalized while the species is left lowercase.

The king eider is in the genus *Somateria*, Greek for "woolly body." The common eider and the spectacled eider are also in this genus. The king eider's species name is *spectabilis*, Latin for "showy" or "spectacular" (which king eiders definitely are). So its full scientific name is *Somateria spectabilis*.

If this cocktail had a scientific name, it would be *Malus autumnus*: "autumn apple." It should hold you over until the winter waterfowl start to arrive!

Fun Bird Fact: In case you're curious, the common eider's species name (coined by Linnaeus himself!) is *mollissima*, meaning "very soft," while the spectacled eider's species name is *fischeri*, in honor of a scientist named Gotthelf Fischer.

Apple Eider

2 oz bourbon
1 oz apple cider
½ oz lemon juice
⅓ oz maple syrup

Combine all ingredients in a shaker with ice. Shake until chilled. Strain into a chilled coupe glass. Garnish with a fan of apple slices on a cocktail pick.

Blood and Sandpiper

Ladies, if you're in search of a strong female role model, look no further than the red-necked phalarope (*Phalaropus lobatus*). This species takes feminism to a whole new level with a full reversal of the traditional roles that male and female birds play. Female red-necked phalaropes are larger and more colorful than males. They compete for mates, defend their nest sites, and let the males incubate the eggs and care for the young. The female doesn't even stick around to meet the kids; once the eggs are laid, she jets off to find another mate, leaving the male to raise the chicks as a single dad.

While this kind of sex role reversal is rare in birds, red-necked phalaropes are not the only ones to do it. Similar biology has been observed in the two other phalarope species, as well as in painted snipes, jacanas, spotted sandpipers, and Eurasian dotterels (among others).

If you're going to be watching shorebirds, you need a nice beachy cocktail. The Blood and Sand, however, is not it. This Prohibition-era Scotch cocktail, made with

Make Myna Double

orange juice, sweet vermouth, and Cherry Heering, is more suited to a fireside than a seaside. This variation is a little more tropical.

Blood and Sandpiper

1 ½ oz blended Scotch
¼ oz sweet vermouth
¼ oz Cherry Heering
½ oz orange juice
½ oz lime juice
¼ oz orgeat
1 dash Angostura bitters

Combine all ingredients in a shaker with ice and shake until chilled. Strain into a large snifter glass filled with crushed ice. Garnish with a lime wheel and a brandied cherry.

Philadelphia Fish Crow Punch

Crows—and other members of their family, Corvidae—are smart. *Very* smart. When it comes to animals, in fact, it's difficult to find one smarter than a corvid. Most of the cognitive abilities once thought to be confined to great apes have now been demonstrated in crows and their relatives, including multi-step problem solving, a theory of mind, self-recognition, episodic memory, and even empathy. Corvids can create and use tools, count up to six, recognize and remember individual human faces, and understand basic physics.

One of the most obviously impressive of these is tool use. Crows are the only non-primates that can make and use multiple types of tools, including hooks, something previously restricted to humans.[24] Additionally, crows can use those tools to solve multi-step puzzles. In one famous example, researchers put a piece of food in a narrow glass case, too far away for a crow to reach. A long stick that could reach it was placed in a cage and a shorter stick was hung from a branch by a string. New Caledonian crows were able to assess the situation and essentially create a plan. They would perch on the branch and retrieve the short stick by pulling up the string and holding it with their feet. They used the short stick to reach the longer stick in the cage and then used the long stick to reach the food. This suggests a level of innovation and complex thought far beyond anything previously attributed to birds.[25]

So why Philadelphia Fish Crow Punch? Punches were all the rage in the eighteenth century. Instead of ordering individual drinks, friends shared a bowl of punch, ladling out cupfuls. This was certainly the case at the Schuylkill Fishing Company, an angling club in Philadelphia that claims to be the oldest continuously active social club in the English-speaking world. Their signature recipe, now known as Philadelphia Fish House Punch, was enjoyed by an impressively prestigious list of members and guests, including George Washington.

The range of the fish crow (*Corvus ossifragus*), the American crow's slightly smaller sibling, extends from Philadelphia into the southeastern United States, where the inspiration for this bourbon-spiked peach tea punch was found. It's made with lemon oleo saccharum, a syrup made from lemon oil and sugar that is often used in traditional punches.

Philadelphia Fish Crow Punch

2 oz bourbon whiskey
1 ½ oz black tea, brewed and cooled
¾ oz lemon juice
1 ½ oz lemon oleo saccharum
½ of a ripe peach, sliced

Combine peach and oleo saccharum in a shaker and muddle. Add remaining ingredients and fill the shaker with ice. Shake until chilled. Fine-strain into a rocks glass filled with ice. Garnish with a lemon wheel.

LEMON OLEO SACCHARUM

2 lemons
½ cup sugar

Peel the zest from the two lemons in strips, taking as little of the white pith as possible. (Juice lemons for this and future cocktails!) Combine the zest in a bowl or mason jar with the sugar and stir. Gently muddle to release the oils from the peel. Stir one more time, then cover the mixture, letting it sit at room temperature overnight. After about twenty-four hours, the sugar should have liquified. If it remains too dry, you can add a small amount of hot water. Strain out the lemon peel pieces and store the oleo saccharum in the fridge.

Queen's Lark Swizzle

Baby birds take sibling rivalry to a whole other level. A large percentage of nestlings and fledglings die before they reach adulthood, so it's important that they get lots of food from their parents—even if they do this at the expense of their brothers and sisters. In the nest, it's every bird for itself.

When a parent arrives, nestlings immediately try to get its attention. They stretch their necks, open their mouths wide, and make begging calls to indicate how hungry they are. If you're just one gaping beak among many, you need a way to stand out. And what better way to do that than some flashy colors or patterns? Some birds have evolved markings *inside* their mouths that are only visible when they're begging for food. They help compel the parent to feed the young, improve visibility in nests with low light, and help parents recognize brood parasites (see Piña Koel-ada, pg. 92). Then they fade away when the chicks grow up.

The estrildid finches are the undisputed winners when it comes to bright and bizarre mouth markings, but they aren't the only birds that have them. The larks (family Alaudidae) are another group with mouth markings as chicks. The chicks of the horned lark (*Eremophila alpestris*), for example, have three distinctive black spots on their tongues which stand out starkly against their pink and yellow gapes when they beg for food.

They're like a target for the parents, saying, "Here I am! Feed me!"

The Queen's Park Swizzle is a tropical cocktail hailing from Trinidad. Since horned larks aren't found that far south, I've swapped in two exclusively North American ingredients: bourbon and maple syrup. This drink is built in a tall glass with crushed ice and then "swizzled" with a swizzle stick or barspoon until it's perfectly chilled.

Queen's Lark Swizzle

2 oz bourbon
1 oz lemon juice
¾ oz maple syrup
8–10 mint leaves
4 dashes Angostura bitters

Combine mint leaves and maple syrup in the bottom of a highball glass. Muddle gently. Add bourbon and lemon juice and then fill the glass with crushed ice. Insert a swizzle stick or barspoon and spin it by rolling it back and forth in the palms of your hands until the drink is chilled and the glass is nice and frosty. Garnish with the Angostura bitters and a bouquet of fresh mint.

Rob Royal Flycatcher

The royal flycatcher (genus *Onychorhynchus*) is one of the most striking birds on the planet. When you first see it, you may be confused as to why. It's a very plain little bird with a brown back and a tawny belly. You may see the hint of a crest or a small pop of red on the top of its head. But when it is trying to attract a mate or intimidate another male, it transforms. It raises its spectacular crest, a bright and shimmering fan of red, black, and blue. There is nothing else quite like it in the world of birds.

Many have pointed out the similarity between the royal flycatcher's crest and an Aztec headdress. It's appropriate for a bird that makes its home in Central and South America. Since we have the Aztecs to thank for chocolate, I dressed up this Rob Roy with a bit of crème de cacao, a chocolate liqueur. It's only a small adjustment to the original recipe, but like the royal flycatcher's crest, it's hiding in there, waiting to make a big impact.

Rob Royal Flycatcher

2 oz blended Scotch
¾ oz sweet vermouth
¼ oz crème de cacao
1 dash orange bitters
1 dash Angostura bitters

Combine all ingredients in a mixing glass with ice. Stir until chilled. Strain into a chilled cocktail glass. Garnish with a brandied cherry.

Fun Bird Fact: The taxonomy of the royal flycatcher is the subject of some debate. In 2024, the International Ornithological Congress (IOC) reduced the number of recognized species from four to two, whereas the American Ornithological Society places all royal flycatchers in a single species. I chose to discuss them at the genus level until everyone sits down for a cocktail and makes up their mind.

Rusty Rail

When we think about flightless birds, there are some typical examples that come to mind: ostriches, emus, penguins, kiwis...but when it comes to flightlessness, no group of birds can rival the rails (family Rallidae). Of the 134 living species of rails, thirty-two cannot fly.[26] Most of these species evolved flightlessness independently, meaning it has happened dozens of times in this one group, usually on remote islands. On the island of Aldabra in the Seychelles, it has happened not once, but twice.

Fossil evidence shows that a flightless rail in the genus *Dryolimnas* lived on Aldabra over 136,000 years ago. It was wiped out, along with Aldabra's other resident animals, when sea levels rose and covered the island. The water eventually receded and new animals began to colonize Aldabra—including another species of rail. After flying to the island, this rail also lost the ability to fly, evolving into what we know today as the Aldabra rail (*Dryolimnas cuvieri aldabranus*).[27]

The independent evolution of similar traits in related organisms is known as **parallel evolution.** The situation on Aldabra is even more unusual, however, because the parallel evolution happened in the *same place* at different *times*. This unusual situation is called **iterative evolution.** It's interesting enough that many news outlets picked up the story.

However, most of them got it wrong. Even *Smithsonian* magazine's headline was "How evolution brought a flightless bird back from extinction."[28] That's unfortunately just not accurate. Because it evolved independently, the Aldabra rail is not the same bird as the extinct *Dryolimnas* rail that lived on Aldabra before it. But saying so gets more clicks than "How iterative evolution caused an extinct flightless bird to eventually be replaced by a very similar one." (It's also far more concise.) More discussion of this gets to the question of what a "species" really is, and that's a topic that deserves a whole book to itself—probably one without cocktail recipes, unfortunately.

The Rusty Nail is a simple, two-ingredient cocktail that was particularly popular in the '60s and '70s. This Rusty Rail keeps the Scotch and Drambuie (a Scotch liqueur flavored with honey, herbs, and spices) of the original, but turns it into a far more crushable sour, the kind you might enjoy while vacationing in the Seychelles.

Rusty Rail

2 oz Scotch
½ oz Drambuie
1 oz lemon juice
½ oz honey syrup (pg. 37)
1 egg white

Lightly whip your egg white with a whisk to break it up. Add it to a shaker with all remaining ingredients. Fill the shaker with ice and shake very well. Strain into a rocks glass filled with ice and garnish with a lemon wheel.

Ingredient Note: Egg whites are a classic ingredient in many sour cocktails, but there are always risks to consuming raw eggs. Make sure your eggs are very fresh. Alternatively, you can use a foamer like Fee Foam or substitute 3 tablespoons of the liquid from a can of chickpeas (known as aquafaba).

Sage-Grouse

A male greater sage-grouse (*Centrocercus urophasianus*) performing a strut display is a bizarre, alien sight. It alters its posture, standing erect with its tail fanning out and its wings spreading forward. It swishes its wings and bobs its head, bouncing its neck pouch so that its two bulbous yellow air sacs hit together with a hollow popping sound, all while making a cooing, gurgling call. Dozens of males will gather together to perform this ritual with the goal of attracting a mate.

When male birds display for females in a group, it's called a **lek.** Lekking certainly makes things convenient; the females know just where to go to get their pick of mates. Males in lekking species don't contribute to parental care, so each male wants to mate with as many females as possible. The females want to find the healthiest, strongest male with the best display, and they usually agree on which one this is. A single alpha male will get the vast majority of mating opportunities.

Lekking sage grouse arrange themselves so that the highest-ranking male is in the center of the lek, and proximity to him indicates quality and status. Even though the majority of the males won't be chosen by a female, they still benefit from participating in the lek. Young males are able to learn from the displays of experienced males. Lower-ranking males near the alpha male may also attempt to "sneak" chances to mate with the females. And if the lower-ranking males are related to the alpha male, then helping him reproduce is still in their evolutionary best interest. It really brings new meaning to the term "wing man!"

This sage grouse-inspired recipe is a type of smash, a shaken cocktail flavored with muddled fruit or herbs—in this case, apple and (of course) fresh sage.

Sage-Grouse

- 1 ½ oz rye whiskey
- ¾ oz lemon juice
- ½ oz honey syrup (pg. 37)
- ¼ of an apple, sliced
- 5 sage leaves
- ¼ oz peated Scotch, to top

Combine apple, sage, and honey syrup in the bottom of a shaker. Muddle well to release the apple's juice. Add whiskey and lemon juice. Fill the shaker with ice and shake until chilled. Strain into a rocks glass filled with crushed ice. If you like your drink a bit smoky, drizzle the peated Scotch on top. Garnish with apple slices and a fresh sage leaf.

Scarlet Rye-bis

If you lined up all of the species of ibis in one place, the scarlet ibis (*Eudocimus ruber*) would stand out. While its relatives have feathers in shades of brown, white, and black with a few pops of iridescence and red here or there, the scarlet ibis opted for a vivid pink. What gives its plumage that distinctive color?

Bird coloration comes from two things: the microscopic structure of feathers, and what pigment molecules are present. There are three types of pigments involved in bird coloration: melanins, which make feathers black or brown; carotenoids, which make reds, oranges, and yellows; and porphyrins, which usually make pinks, browns, and reds. Blue coloration and iridescence come from feather structure, which affects the way the feathers reflect light. Some colors are the result of a combination of these two things. For example, most green coloration is caused by yellow carotenoid pigments in structurally blue feathers.

The vivid red-pink of the scarlet ibis comes from carotenoids. These pigments are unique because they are the only ones that birds cannot produce themselves. Instead, they must obtain them from their diet. Like flamingos and the roseate spoonbill, the scarlet ibis gets its carotenoids from the tiny shellfish it eats. Captive birds need to be given supplemental beta-carotene or they will lose their coloration. The richer their diet is in carotenoids, the more vivid their plumage will be. For more on this, see the Booby Trap (pg. 56).

The Scarlet Rye-bis gets its red coloration from Martini Fiero, a bitter red vermouth. If you don't have it, try Aperol instead.

Fun Bird Fact: The turacos (order Musophagiformes) produce a special porphyrin pigment called turacoverdin that makes their feathers green. They are the only birds that have it.

Scarlet Rye-bis

1 ¼ oz rye whiskey
1 oz Martini Fiero
1 tsp sweet vermouth
1 dash Peychaud's bitters
1 dash Bittercube Cherry Bark Vanilla bitters (or another woodsy bitter)

Combine all ingredients in a mixing glass with ice. Stir until chilled. Strain into a rocks glass over one large ice cube. Garnish with an orange twist (discarded).

Other Base Spirits

Bamboo Partridge

There are some birds you don't see until it's too late. You hear the flutter of wingbeats and catch a quick glimpse of a partridge, quail, or rail flying away. Sometimes it's the only way to see them!

Taking off that quickly isn't easy, and these birds have wings that are specially designed for it. Their short, rounded wings allow them to take off rapidly, fly very fast for short periods of time, and easily maneuver around obstacles like trees. Their flight involves lots of flapping, using the large surface area of their wings to deflect air and create speed and lift. It's very effective, but not very efficient for long periods of time. For this reason, birds that spend more of their time in the air than on the ground don't have wings like this. Soaring birds like vultures or hawks have long, broad wings that can catch warm air currents and keep them aloft. Fast fliers like swifts and falcons have narrow, tapered wings to reduce drag and cut through the air.

What can we tell about the Chinese bamboo partridge (*Bambusicola thoracicus*) from its wings? They are longer and narrower than those of other partridges. This means that the Chinese bamboo partridge can fly for longer distances than its relatives, while still weaving around trees and other obstacles. It often moves around the forest on the wing rather than the ground.

What can you tell about this Chinese-inspired variation of the Bamboo cocktail just by looking at it? A base of sherry and vermouth instead of a higher-proof spirit means that it will have a lower ABV, making it a good choice tonight if you're planning to get up early to do some birding tomorrow!

Bamboo Partridge

1 ½ oz amontillado sherry
1 ½ oz dry vermouth
1 tsp five spice oolong syrup
1 dash orange bitters

Combine all ingredients in a mixing glass with ice and stir until chilled. Strain into a chilled coupe glass and garnish with a lemon twist (discarded).

FIVE SPICE OOLONG SYRUP

½ cup sugar
½ cup brewed oolong tea
2 star anise
½ tsp fennel seed
½ tsp peppercorns (white or Szechuan)
¼ tsp cloves
1 cinnamon stick, smashed

Combine sugar and tea in a saucepan over medium heat and bring to a simmer, stirring frequently. Stir in the five spices and let simmer for about one minute. Remove from heat and let cool completely. Strain through a fine mesh strainer. Store the syrup sealed in the refrigerator.

Bulbullini

It's pretty tough to discover a new bird species these days. In the eighteenth and nineteenth centuries, naturalists were describing unique birds left and right as the Western world explored and expanded. Today, when we do occasionally find a new species of bird, it is usually hiding in plain sight.

One recent example of this is the cream-eyed bulbul (*Pycnonotus pseudosimplex*), which was described by ornithologists from Louisiana State University in 2019. For decades, it was thought that cream-vented bulbuls (*Pycnonotus simplex*) in Borneo could have one of two iris colors: red or cream. But when researchers sequenced the DNA of these birds, they found that the red-eyed birds and the cream-eyed birds weren't closely related at all. They were two entirely different species. Since the red-eyed Bornean birds were most closely related to the other cream-vented bulbuls in southeast Asia, the cream-eyed species was deemed the new one.

Okay, you've discovered a new bird species—what happens next? The LSU scientists published a paper in the *Bulletin of the British Ornithology Club* explaining why the cream-eyed bulbul should be considered a separate species.[29] They described its appearance and size, gave it a new common and scientific name, and designated a **type specimen.** Type specimens are the definitive references for what a species or subspecies looks like. In birds, type specimens are usually stuffed

and dried skins that preserve the general shape and color of a bird. These specimens are treated with special care in museum collections, as they are considered important references. The type specimen for the cream-eyed bulbul now sits in a museum drawer at LSU, available for any researcher to reference.

The type specimen for the classic Bellini is a delightfully simple mix of peach puree and sparkling wine. A drink so easy to make and yet so delicious is a rare thing, so I will not disrespect it by making this one too elaborate. Swap the peach puree for mango juice, and you have a new species of cocktail with a tropical, Bornean bulbul-approved twist.

Bulbullini

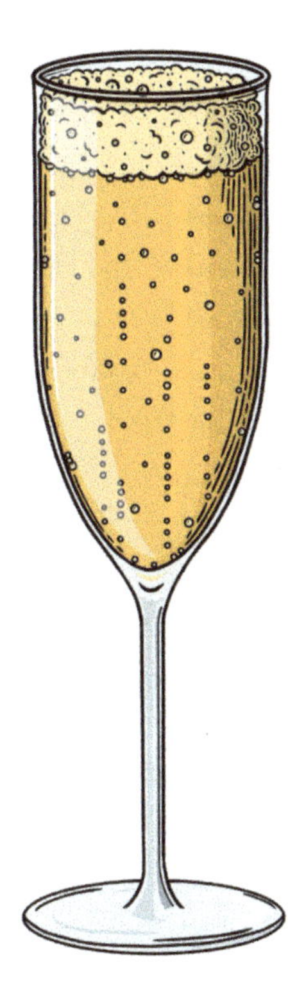

2 oz mango juice
4 oz sparkling wine

Pour mango juice into a champagne flute and top with sparkling wine.

Get Rheal

Though they may be flightless, the ratites certainly know how to get around. Africa has the ostriches, Australia has the emu and the cassowaries, and South America has the rheas (order Rheiformes). Not as well-known or quite as large as their cousins, rheas are no less striking. They can be found across the southern half of South America, which is split between the two rhea species: the greater rhea (*R. americana*) in the east and the lesser or Darwin's rhea (*R. pennata*) in the west.

Why is it that every continent in the Southern Hemisphere seems to have a large flightless bird? Scientists used to think this meant that all these birds came from a single flightless ancestor that lived in Gondwana, the southern supercontinent, and that the ancestors of today's ratites rode on the different pieces as they broke apart. But DNA sequencing has revealed that the ratites lost the ability to fly at least three times in their family tree.[30] It seems like, if there's room for a large flightless bird on a continent, one will evolve there!

What's more, the tiny, flighted tinamous of South America were found to be part of this otherwise flightless group. Bird genetics has certainly been full of surprises!

For more on how often flightlessness evolves, check out the Rusty Rail (pg. 167).

This refreshing spritz is made with Trakal, a unique Patagonian spirit distilled from apples, pears, and crabapples. If you're looking for a substitute, the closest thing would be a dry fruit eau-de-vie.

Get Rheal

1 oz Trakal
½ oz pear liqueur
2 oz sparkling wine
1 ½ oz tonic water
1 sprig fresh rosemary
1 lemon wheel

Fill a large wine glass with ice. Add a sprig of fresh rosemary and a lemon wheel. Pour in Trakal and pear liqueur, then top with tonic and sparkling wine. Stir very briefly and garnish with a lemon twist (discarded) and another sprig of rosemary.

Hoatzinfandel

The hoatzin (*Opisthocomus hoatzin*) of the Amazon rainforest may genuinely be the weirdest bird in the world. There is nothing else quite like it. In fact, it is the only bird to have its own order, Opisthocomiformes. And scientists can't agree on where to place it in the avian family tree. We don't know what its closest living relatives are.

The hoatzin has an incredibly unusual digestive system. It primarily eats leaves, which are digested and fermented by bacteria in the hoatzin's massive crop before being passed to its proportionally tiny stomach. This is more similar to mammalian ruminants like cows than to other birds. As a result of all that fermenting vegetation, the hoatzin does not smell very pleasant, and it is colloquially known as the "stinkbird."

Hoatzins nest in trees over rivers and lakes, and their young are born with dinosaur-like claws on their wings that help them hold on to branches and climb through the trees until they can fly. If they do fall into the water, they are able to swim and cling to vegetation, climbing back to their nesting area.

Seeing wild hoatzins in Peru was one of my top birding moments, so it seems only right to make a Pisco Sour in their honor. This one has a vividly colored float of Zinfandel, a fruity, full-bodied red wine.

Hoatzinfandel
OLD VINE
2025

Hoatzinfandel

2 oz Pisco
½ oz lime juice
½ oz lemon juice
¾ oz passion fruit syrup (pg. 38)
1 egg white
¾ oz Zinfandel wine
1 dash Angostura bitters

Lightly whip the egg white with a whisk to break it up. Add it to a shaker with pisco, lime juice, lemon juice, and passion fruit syrup. Shake until chilled and frothy. Strain into a rocks glass filled with ice. Slowly top with the Zinfandel by pouring it down the back of a spoon. Dash bitters on top.

Jacamartinez

The rufous-tailed jacamar (*Galbula ruficauda*) of South and Central America is a small emerald- and copper-colored bird with a long tail and bill. It's a specialized insectivore, and butterflies are one of its favorite foods. Unfortunately for the jacamar, some of the butterflies within its range taste absolutely terrible. This is no accident; tasting bad is a great way for a butterfly to avoid being eaten. But being caught and spit out is still quite traumatic, as is getting a beakful of toxic butterfly. It's better for both the predator and the prey if the butterfly can let the jacamar know ahead of time just how bad it's going to taste.

The *Heliconius* butterflies that the rufous-tailed jacamar likes to hunt do just that. These butterflies have distinct patterns that are easy to recognize—even for a bird. When the jacamar is young, it will inevitably catch a bad-tasting butterfly. It will spit the offender out and quickly learn from the mistake, avoiding butterflies with similar wing patterning from then on.

The Martinez is a predecessor of the Martini, harkening back to a time when people didn't like their drinks quite so dry. This one swaps out Old Tom Gin for an aged cachaça, a Brazilian spirit made from sugar cane juice, and adds a bit of tropical banana flavor in honor of this lovely neotropical bird. No warning coloration needed here—the Jacamartinez is perfectly palatable.

Jacamartinez

2 oz aged cachaça
1 oz sweet vermouth
¼ oz banana liqueur
1 dash Angostura bitters

Combine all ingredients in a mixing glass. Add ice and stir until chilled. Strain into a chilled cocktail glass. No garnish.

Ingredient Note: If you don't have aged cachaça, an aged rhum agricole is the next best choice.

Jaeger Bomb

In this book you'll find birds that lie (Plover Club, pg. 227) and cheat (brood parasites, pgs. 60, 92, and 114), so it's no surprise that there are also birds who steal. They do it so often, in fact, that their way of life has a name: **kleptoparasitism.**

Some birds will occasionally indulge in kleptoparasitism, but there are many seabirds who specialize in it, including the frigatebirds and the skuas. One species of skua, the parasitic jaeger (*Stercorarius parasiticus*), even gets its name from the practice. Parasitic jaegers make their living stealing the fish and seafood caught by other seabirds, chasing and dive-bombing their victims until they drop the goods. They frequently gang up on their targets, sometimes taking on birds much larger than themselves.

There is no drink more fitting for a dive-bombing jaeger than that college campus mainstay, the Jäger Bomb. A shot of Jägermeister dropped into a glass of Red Bull is not exactly a craft cocktail, but the German digestif shouldn't be written off entirely because of one bad drink. Try it in this Margarita-style recipe that still incorporates the unique flavor of Red Bull as a syrup.

Jaeger Bomb

2 oz Jägermeister
¼ oz triple sec
1 oz lime juice
¾ oz Red Bull syrup

Combine all ingredients in a shaker with ice and shake until chilled. Strain into a chilled coupe. No garnish.

RED BULL SYRUP

1 can Red Bull
1 cup sugar

Combine the Red Bull and sugar in a saucepan over medium heat. Stir until sugar is dissolved. Let cool completely before using. Store in the refrigerator.

Kite-Pirinha

The brahminy kite (*Haliastur indus*) is a striking bird of prey found in India, southeast Asia, and Australia. It is a particularly lovely bird with a white hood, a reddish-brown body, and black wing tips. And while they may be the least notable thing about the brahminy kite's beautiful plumage, it's those wing tips I want to talk about.

If you're an avid birdwatcher, or even a casual one, you may have noticed that most birds of prey have black tips on their primary feathers. In fact, many birds do: gulls, terns, storks, geese...the list goes on. What makes it such a popular feature?

Black feathers are black because they contain a pigment called **melanin.** Melanin doesn't just affect the color of feathers—it also makes them stronger and more durable, protecting them from wear and tear. And those primary feathers are exposed to a lot of wear and tear when birds spend most of their time flying. Having extra melanin helps them keep their shape and stay in good condition.

The Caipirinha is a Brazilian cocktail made with the country's unique rum-like spirit, cachaça. This Kite-Pirinha takes inspiration from an Indian dish called guava chaat. It's often served by street vendors in some of the areas where brahminy kites are found.

Kite-Pirinha

2 oz cachaça
½ oz guava nectar
½ oz sugar
1 lime, cut into eighths
Chili powder or masala chaat, for rim

Rim half a rocks glass with chili powder or masala. Combine lime and sugar in the glass and muddle well to release the lime juice. Fill the glass with ice, then add guava nectar and cachaça. Stir briefly.

LBB

Birders have a lot of their own lingo. "Twitchers" are birders who go to great lengths to chase down rare birds. A "spark bird" is the bird that inspired a birder to begin their hobby. A "life bird" or "lifer" is a new bird seen for the first time. "Jizz" refers to the general appearance and behavior of a bird. And "LBB" is an acronym standing for "little brown bird," used to refer to the many plain brownish songbirds that can be so difficult to distinguish in the field (alternatively, LBJ, "little brown job," is used).

The dunnock (*Prunella modularis*) is sort of the original LBB. Native to Europe and western Asia, its name literally means "little brown thing" in Old English (*dun* meaning brown and *-ock* being a suffix added to small things, as in *hillock* and even *buttock*). These birds are so unremarkable and easily overlooked that they were able to hide some fascinating—and downright salacious—bird behavior in plain sight.

Prudish British ornithologists always thought that dunnocks were monogamous. But when Nick Davies began studying them in the 1980s, he was stunned to learn that this wasn't the case. In fact, these unassuming little garden birds engage in all sorts of cheating and polygamy. Male dunnocks will even peck at a promiscuous female's cloaca before mating with her to induce her to eject any sperm that is already inside.[31]

Even with this precaution, male dunnocks frequently end up caring for chicks that aren't their own.

This cocktail is an updated version of the B&B, a retro drink made from equal parts brandy and Benedictine, a French herbal liqueur. Adjusting the proportions makes it a much better cocktail, and the addition of a little Licor 43 introduces notes of vanilla, citrus, and cinnamon. It's definitely still on the sweet side, so save this one for after dinner!

LBB

2 oz brandy or Cognac
1 oz Benedictine
½ oz Licor 43

Combine all ingredients in a mixing glass. Add ice and stir until chilled. Strain into a brandy snifter containing one large ice cube. Garnish with an orange twist.

Negrito Negroni

I distinctly remember the first time I saw an Andean negrito (*Lessonia oreas*). I was on the Altiplano in Peru, high in the Andes, peering at the birds gathered near a small lake through my binoculars. Among the waterfowl, a little black and brown bird perched on a stone. It was the austral winter; the cold wind stung at my cheeks and my fingers eventually went numb. You could see the breeze ruffle the negrito's feathers, but he didn't seem bothered by the temperature.

Feathers really are amazing things. They keep cold-climate birds warm, provide buoyancy to waterbirds, give male birds something to show off, and—of course—allow birds to fly. They are essentially modified scales, made of the same stuff as horns, hair, and fingernails: keratin. But their structure is far more complex than any of these others. They have a center shaft called a **rachis** with hundreds of **barbs** protruding from it. These are latched together by tiny hooks called **barbules.** When birds run their beaks along their feathers as they preen, they are "zipping" their feathers back together, interlocking these barbules. Indeed, birds spend a lot of time maintaining their feathers: cleaning them, arranging them, removing parasites, and covering them with oil from a gland near their tail called the **uropygial gland.** The negrito's quick sallies from its low perch in the frigid wind may have looked effortless, but it takes a lot of work—and millions of years of evolution—to be able to do what it does every day.

On that same trip to Peru, I drank a lot of chicha morada, a spiced juice made from purple corn and pineapple. Here those same flavors are infused into a Pisco Negroni in the Andean negrito's honor.

Negrito Negroni

1 oz purple corn-infused Pisco
¾ oz Luxardo Bitter Bianco
¾ oz pineapple-infused blanc vermouth
¼ oz allspice dram

Combine all ingredients in a mixing glass with ice and stir until chilled. Strain into a rocks glass over one large ice cube. Garnish with a slice of dried pineapple.

PURPLE CORN-INFUSED PISCO

1 cup Pisco
¼ cup maíz morado

Combine pisco and purple corn in a mason jar or other container. Shake or swirl briefly and let sit overnight at room temperature. After twenty-four hours, strain out the corn.

PINEAPPLE-INFUSED BLANC VERMOUTH

1 cup blanc vermouth
½ cup fresh pineapple chunks (or 1 cup pineapple rinds—a great way to reduce waste)

Combine vermouth and pineapple in a mason jar or other container. Shake or swirl briefly and let sit overnight in the refrigerator. After twenty-four hours, fine-strain to remove the pineapple. Store in the refrigerator.

Pitohui Punch

In 1988, graduate student Jack Dumbacher was mist-netting birds in Papua New Guinea. He was trying to catch birds of paradise, but every now and then, a hooded pitohui (*Pitohui dichrous*) would get caught in the net. While releasing one, a cut on Dumbacher's finger began to burn and tingle, so he absently popped it into his mouth. Strangely, his mouth, too, began to burn and go numb like he'd eaten something spicy.

The sensation eventually went away, and as he was busy with work in a jungle full of unusual plants and insects, Dumbacher didn't give the occurrence much thought—until he returned to New Guinea a year later, and a high school volunteer commented that something similar had happened to him after handling a hooded pitohui. The next time he caught one, Dumbacher removed a feather from the bird and put it on his tongue. He immediately felt the same sensation again and knew that he had stumbled upon something utterly novel in the world of ornithology: a poisonous bird.

It turns out that the feathers of the hooded pitohui are laced with homobatrachotoxin, the same chemical found in the skin of some poison dart frogs. They likely obtain it from beetles that they eat. It may deter predators, protect feathers against parasites, or both. This toxic trait has now been found in other Papuan bird species, including other pitohuis, the blue-capped ifrit, the regent whistler, and the rufous-naped bellbird.

This Pitohui Punch is a clarified milk punch, a type of cocktail that has been made for hundreds of years. The acidic ingredients of the drink are combined with milk, causing it to curdle. When the milk fats are filtered out, the resulting cocktail is clear, silky-textured, and has a refrigerator shelf life of up to a year. This milk punch is additionally infused with buzz buttons (also called Sichuan buttons), which give it a subtle tongue-numbing effect not unlike that of a pitohui's feathers.

Pitohui Punch

(2 servings)

- 3 oz Batavia arrack
- 2 oz dry vermouth
- 2 oz ginger lemon tea, brewed and cooled
- 2 oz lime juice
- 1 ½ oz ginger syrup (pg. 37)
- 1 ½ oz pineapple juice
- 2 oz whole milk
- 6 buzz button flowers

Combine arrack, dry vermouth, tea, juices, and syrup in a mixing glass or cup measure. Remove the buzz button petals from the flowers and stir them into this mixture. Pour whole milk into another container. Pour the cocktail and buzz button mixture into the milk. Do not stir. Let sit in the fridge for several hours. Fine-strain, discarding curds, and then strain through a coffee filter. Serve in a rocks glass over one large ice cube, garnished with a buzz button flower.

Rosélla

Parrots like the crimson rosella (*Platycercus elegans*) are known and loved for their spectacularly colored plumage. But when it comes to bird feathers...we don't see the half of it. Well, a quarter of it.

Animals see color using special cells in their retinas called **cone cells.** Humans have three types of cones, each one allowing us to see a different range of colors. People who are colorblind lack one or more of these cone types. Birds, on the other hand, have *four* types of cones: three that allow them to see what we see, and one that lets them perceive ultraviolet (UV) light.

This means that when we look at a bird like a crimson rosella, we're not seeing what other birds see at all. Rosellas have patches of feathers with colors in the UV spectrum, colors it's hard for us to even imagine. But we can see where these patches are by looking at a bird under UV light. They will reflect it and appear to glow.

Cocktails can glow under UV light as well! By adding a little bit of riboflavin (also known as vitamin B2), you can give your drink a greenish glow under black light. It works well in this delicious and floral rosé sangria. If you're not in the mood for an avian retina-inspired gimmick, add a few fresh strawberries to your glass instead.

Rosélla

3 oz rosé wine
½ oz elderflower liqueur
¾ oz lemon juice
½ oz strawberry syrup
¼ oz liquid riboflavin

In a stemless wine glass or rocks glass, combine elderflower liqueur, lemon juice, strawberry syrup, and riboflavin. Stir. Fill the glass with ice and top with the rosé.

Ingredient Note: If you have capsules of riboflavin/vitamin B2, dissolve 100 mg of the powder inside the capsules in 500 ml (about 3 cups) water.

STRAWBERRY SYRUP

6 strawberries, sliced
½ cup sugar
¼ cup water

Combine sugar and water in a saucepan over medium heat. Bring to a simmer, stirring until sugar is dissolved. Add strawberries and continue to gently simmer until they soften, about three minutes. Remove from heat. Use the back of your spoon or a muddler to smash the berries into the syrup. Cover and let cool. Fine-strain to remove the strawberries (which make a great topping for ice cream!) and store in the refrigerator. Try it in a Gimlet, Sangria, or nonalcoholic lemonade.

Sidecardinal

Even someone who has never picked up a field guide can probably identify the northern cardinal (*Cardinalis cardinalis*). Adult male cardinals are a striking shade of red, making them some of the most colorful birds that commonly visit our feeders. Female cardinals, on the other hand, are quite plain. Their plumage is brown, with only small patches of red to hint at the color of their spectacular male counterparts. Why?

Sexual dimorphism is when males and females of a species look different. Cardinals are a perfect example of how this usually manifests in birds: a drab, nondescript female and a bright and flamboyant male. In these species, the females choose which male they want to mate with, so the males have evolved these bright and impressive signals to attract their attention. It's not just about catching a girl's eye—a male has to be healthy to produce bright plumage or elaborate feathers. He must also evade predators despite those cumbersome plumes and "here I am!" colors. If he can do these things, it indicates to females that he will be a quality mate who will pass these same traits on to her offspring. Meanwhile, the female sticks with a sensible shade of brown or grey. Men, am I right?

Like the northern cardinal, this Sidecardinal cocktail comes in two colors. Both versions start with the recipe for a classic Sidecar cocktail: Cognac, orange liqueur, and lemon juice, served with a sugared rim. From there, you can be as choosy as a female cardinal.

Sidecardinal (*"Male" Version*)

1 ½ oz Cognac
¾ oz triple sec
¾ oz lemon juice
½ oz cranberry juice

First, prepare your glass. Fill a shallow dish with sugar and (optional) dried cranberry powder. Run a lemon wedge along the rim of a coupe and dip it in the dish to rim it with sugar. Combine all the cocktail ingredients in a shaker with ice and shake until chilled. Strain into your prepared glass.

Sidecardinal (*"Female" Version*)

1 ½ oz Cognac
¾ oz lemon juice
½ oz triple sec
½ oz Amaretto

Prepare your glass by running a lemon wedge along the rim of a coupe and dipping it in sugar. Combine all the cocktail ingredients in a shaker with ice and shake until chilled. Strain into your prepared glass. Garnish with a brandied cherry.

Suffering Bustard

If there were a *Guinness Book of Bird Records*, you can imagine what some of the entries would be. Largest bird: common ostrich. Smallest bird: bee hummingbird. Largest egg: common ostrich again. Largest egg relative to body size: North Island brown kiwi. Fastest bird: peregrine falcon. Highest-flying bird: Rüppell's vulture. Finally, the heaviest flying bird: the kori bustard (*Ardeotis kori*).

The number one rule of flying is that you have to be light. Every single aspect of a flying bird's anatomy and physiology is optimized to reduce mass, so holding this particular record is no mean feat. Male kori bustards can weigh as much as *forty pounds*. For context, a sandhill crane (which I think of as a pretty big flying bird!) weighs about ten pounds. Condors are the only flying birds that even come close to the kori bustard.

This drink is based on a World War II-era cocktail called the Suffering Bastard, so named for its reputation as a hangover cure. It was adopted by Tiki bartender Trader Vic, who would famously serve it in a mug that looked like a classic Tiki figure nursing a headache. This is a split-base cocktail, originally made with a mix of Cognac and gin (here, Cognac and rye). Pamplemousse grapefruit liqueur pays tribute to a popular soda made in Cameroon, one of the African countries the kori bustard calls home.

Suffering Bustard

1 oz rye
1 oz Cognac
¾ oz pamplemousse liqueur
¾ oz lime juice
2 oz ginger beer

Combine rye, Cognac, lime, and pamplemousse liqueur in a shaker. Add ice and shake until chilled. Strain into a Tiki mug filled with ice. Top with ginger beer and garnish with a grapefruit slice and a mint bouquet.

Mocktails

Bella Loona

Birds exhibit a huge array of behaviors when it comes to mating and raising chicks. We've talked a little about birds that sneak around (LBB, pg. 194), male birds that mate with many females (Sage-Grouse, pg. 170), and female birds that mate with many males (Blood and Sandpiper, pg. 155). But what ever happened to good, old-fashioned monogamy? It's alive and well in the avian world—sort of.

Monogamy means different things to different birds. Some birds mate for life. Seasonally monogamous birds stick with one mate for a season, but will re-pair the next year. Socially monogamous birds act like a pair for the purposes of raising young, but may engage in extra-pair copulations. And for some, like the common loon (*Gavia immer*), it's more about the real estate.

Loons return to the same breeding site year after year. The males arrive first to establish the territory, and the females join a few days later. Pairs stay together for years. But if a different male takes over the territory, the female will breed with him instead. If the pair is evicted from the territory, they may both re-pair with new mates. Both of them are more loyal to the nesting site than to each other. This is considered **serial monogamy**—mating for multiple seasons but not necessarily for life.

When the moon hits your eye...it may not be amore for loons, but you're sure to fall in love with this nonalcoholic version of the Bella Luna cocktail. It has all the flavors of the original with none of the booze!

Bella Loona

1 oz lemon juice
½ oz violet syrup
5 oz elderflower soda

Fill a highball glass with ice. Add lemon juice and violet syrup and stir briefly. Top with elderflower soda. Garnish with a lemon wheel.

Compulsive Lyrebird

The name of this cocktail is more than just a pun—it's also an apt description. The superb lyrebird (*Menura novaehollandiae*) of Australia may get its name from its lyre-shaped tail feathers, but it's also a compulsive liar. It's one of the most skilled and convincing avian mimics. Superb lyrebirds can imitate dozens of other species, and even human-made noises like car alarms or jackhammers.

Why do lyrebirds—and other master mimics like mockingbirds, drongos, and starlings—perform these imitations? You may be surprised to learn that we don't really know. Perhaps female birds are impressed by males with big repertoires, so they pick up whatever songs and sounds they can. Perhaps a male imitating several other bird species is trying to keep those species away from his territory, leaving him with more resources. Or maybe it's more of a biological accident.

Most birds learn their songs from their parents and neighbors, the same way human children learn to talk. Their brains are already programmed to learn and repeat what they hear, and some species are predisposed to just keep doing it.

Even if you've never been tricked by a convincing lyrebird, you *might* be fooled by this mocktail. As bitter and refreshing as an Aperol Spritz but with none of the alcohol, it's the kind of mimic we can all get behind. Look for Lyre's Aperitif Rosso—Lyre's nonalcoholic spirits are named for the lyrebird!

Compulsive Lyrebird

2 oz nonalcoholic aperitivo
3 oz nonalcoholic sparkling wine
1 oz club soda
3 strawberries, cut in halves

Place strawberries into a wine glass and fill with ice. Add remaining ingredients and stir briefly.

Culture Vulture

Protecting endangered species is about more than just saving pretty birds. The fact is that ecosystems—even those that humans are already a major part of—are delicate networks with millions of interconnected parts. Removing a single one can have major unexpected consequences. The three Indian vultures in the genus *Gyps* are a perfect example of this fact. When these once-abundant species became critically endangered, over half a million people died.

In the early 1990s, the Indian patent on a veterinary painkiller called diclofenac expired, resulting in the widespread use of the drug to treat cattle. Unfortunately, diclofenac turned out to be fatal to the vultures that fed on dead cattle. This resulted in a massive decline in India's vulture population, from the millions to the thousands. White-rumped vulture (*Gyps bengalensis*, pictured) numbers declined by *99.9%*.[32]

For bird and nature lovers, this is enough of a tragedy in itself. But it also started a chain reaction of events with devastating consequences. Without vultures to feed on dead animals, the carcasses were left to rot, resulting in the spread of deadly bacteria and infections through drinking water. Those that didn't rot were consumed by feral dogs, the number of which multiplied rapidly as a result of the new food source. Because of this, rabies infections spiked as well. Researchers estimate that 500,000 people died as a result of the vulture decline between 2000 and 2005 alone.[33]

Diclofenac was banned in 2006, but vulture numbers have not rebounded. There is emerging evidence that a new anti-inflammatory drug, nimesulide, is also killing vultures.[34] Will we learn from our past mistakes? Only time will tell.

Culture Vulture

2 oz chai tea, brewed and cooled
1 oz lemon juice
1 oz mango syrup
1 tsp plain yogurt
½ cup mango chunks
Cardamom, to top

Combine mango chunks and mango syrup in the bottom of a shaker and muddle well. Add tea, lemon juice, and yogurt. Fill the shaker with ice and shake until chilled. Strain into a rocks glass filled with crushed ice. Sprinkle cardamom on top.

MANGO SYRUP

The leftover peels and core from 1 mango or half of a mango, cut into chunks
½ cup sugar
¼ cup water

Combine sugar and water in a saucepan over medium heat. Bring to a simmer, stirring until sugar is dissolved. Add mango chunks or mango scraps. Let simmer about five minutes, stirring occasionally. Gently muddle the mango to release its juice. Cover and remove from heat. Let cool completely and then fine-strain. Store in the refrigerator. Try it in a Daiquiri or Margarita.

Hawkward Position

Look at this text as you're reading it. You're keeping it in the center of your vision, where it's sharp and in focus. Beyond the words you're looking at, your vision becomes blurrier. Objects in the corner of your eye aren't really in focus. This is because of the way your retina is arranged. There is a small, central area called the **fovea** that gives you precise, clear, high-acuity vision, and you use it every time you look directly at something. You have one in each eye. But some birds have *two*.

It's hard to imagine what it would be like to see with two foveae, just as it's impossible to picture the additional colors that birds perceive (see Rosélla, pg. 203). Most birds have their eyes on the sides of their head, so a central fovea like ours gives them sharp vision out to either side. This is helpful if you're a bird that needs to be on the lookout for predators. But if you *are* a predator, like the red-tailed hawk (*Buteo jamaicensis*), that's not quite good enough. So many birds of prey have an additional fovea on the side of their retina that allows them to have sharp vision to the sides *and* in the front. This trait is shared by other birds that need to be able to judge distance and focus in on objects, including kingfishers, terns, hummingbirds, and swallows.

Hawkward Position

2 oz verjus rouge
2 oz grapefruit juice
1 oz spiced cranberry syrup
3 oz club soda
1 sprig fresh tarragon

Add tarragon and cranberry syrup to the bottom of a shaker and muddle gently. Add verjus and grapefruit juice. Fill the shaker with ice and shake until chilled. Strain into a highball glass filled with ice and garnish with a sprig of fresh tarragon and a few fresh cranberries. Serve with a straw.

Ingredient Note: Verjus is a very acidic juice made from unripe grapes. It tastes a bit like a mild vinegar or fruit shrub.

SPICED CRANBERRY SYRUP

1 cup sugar
1 cup water
1 cup cranberries (frozen is fine)
2 cinnamon sticks
1 star anise
8 cloves
Shaved nutmeg

Combine water and sugar in a saucepan over medium heat. Bring to a simmer, stirring frequently until sugar is dissolved. Add cranberries and continue to simmer until they soften and begin to break open, about two minutes. Smash the cranberries with a muddler or the back of your spoon. Add spices and stir. Simmer for one more minute, then remove from heat and cover. Let cool completely before straining out the cranberries and spices. Store in the refrigerator. Try it in an Old-Fashioned or Cosmopolitan.

Lame Duck

One of my favorite topics in ornithology is duck penises.

I love telling people about duck penises, because most people have absolutely no idea that ducks *have* penises, and if that fact alone doesn't shock them, a picture of one certainly will. Duck penises are bizarre, alien-looking organs with prongs and corkscrew turns. They are usually retracted inside the body, but when fully extended they can be quite long—over sixteen inches!

Unfortunately, the reason for this bizarre feature is not nearly so amusing. Male ducks will often mate with females whether they're willing or not. The female ducks would prefer to have some choice in the matter, so they have evolved labyrinthine, spiraled vaginas complete with dead-end turns. In response, male ducks have evolved penises to match, spiraling in the opposite direction so that they can navigate the maze and deposit sperm as close to the ovary as possible. It's what is called

an **evolutionary arms race,** where the female evolves defenses and the male evolves ways to circumvent them.

If I asked you *which* duck you thought might have that sixteen-inch penis, I doubt your first guess would be the tiny, plump lake duck (*Oxyura vittata*). Quite similar to the ruddy duck of North America, this stiff-tailed duck is native to Argentina and Chile. Not only is its penis the longest we know of, it is also covered in spines. The lake ducks use them "like bottle brushes" (that's a direct quote from the prestigious journal *Nature*) to clear out any sperm already in the oviduct before inseminating the female.[35] What can I say? Nature is weird.

This mocktail isn't inspired by duck penises *per se*, but it is a nod to the lake duck's Patagonian home. If you've been to the southern half of South America, you've probably encountered yerba mate, a grassy tea that is traditionally consumed from a gourd with a special straw. Here, it makes a delicious base for a ginger lemonade.

Lame Duck

4 oz yerba mate tea, brewed and cooled
1 oz lemon juice
1 oz ginger syrup (pg. 37)
1 mint bouquet

Fill a highball glass with ice. Add tea, lemon juice, and ginger syrup. Stir briefly and garnish with the mint bouquet.

Plover Club

If birds were eligible for human awards, the competition would be stiff. They'd sweep the Grammys. The Tony for Best Choreography would go to the long-tailed manakin; the satin bowerbird would win for Set Design. At the Olympics, the common ostrich would beat Usain Bolt's top speed, and the gentoo penguin would outswim Michael Phelps. And the Oscar for Best Leading Actor goes to...the piping plover (*Charadrius melodus*).

Piping plovers dig shallow nests on sandy beaches, and while they do their best to camouflage them with grass and shells, they're mostly exposed to predators. So when a threat like a gull, cat, raccoon, or even human approaches the nest, the parent on duty springs into action. They leave the nest (the chicks are camouflaged to blend in with the sand) and begin fluttering helplessly on the ground, one of their wings held out at an awkward angle. The predator, thinking the parent is injured and will be easy prey, follows it away from the chicks. Once the danger is far enough from the nest, the plover will miraculously recover and fly away.

I've singled out the piping plover, but this behavior is common in its genus and, indeed, among birds in general. One study found documented evidence of a total of 285 bird species in fifty-two families performing a broken-wing display.[36] The Oscar race next year may be even more contentious.

A mocktail seems an appropriate choice for a bird that is a master of fakery. This nonalcoholic take on the Clover Club (gin, lemon, raspberry syrup, and egg white) reinterprets the raspberry syrup in the original as a shrub. Also known as drinking vinegars, shrubs are tangy syrups flavored with fruit and vinegar that have their roots in colonial America. This mocktail might not trick you into thinking you're drinking the real thing, but you certainly won't miss the booze!

Plover Club

1 ½ oz raspberry shrub
4 oz club soda

Add shrub to a highball glass filled with ice. Top with club soda and stir briefly. Garnish with a raspberry and a bouquet of fresh mint.

RASPBERRY SHRUB

6 oz fresh raspberries
¾ cup granulated sugar
¼ cup apple cider vinegar

In a bowl, gently mash raspberries. Add sugar and stir well to combine. Transfer this mixture to a jar and allow it to sit, shaking or overturning occasionally, for about twenty-four hours. Once the sugar is dissolved and the mixture is liquid-like, strain it through a fine mesh sieve. Discard the solids. Add apple cider vinegar to the liquid and stir. Adjust vinegar and sugar to taste. Store shrub in the refrigerator.

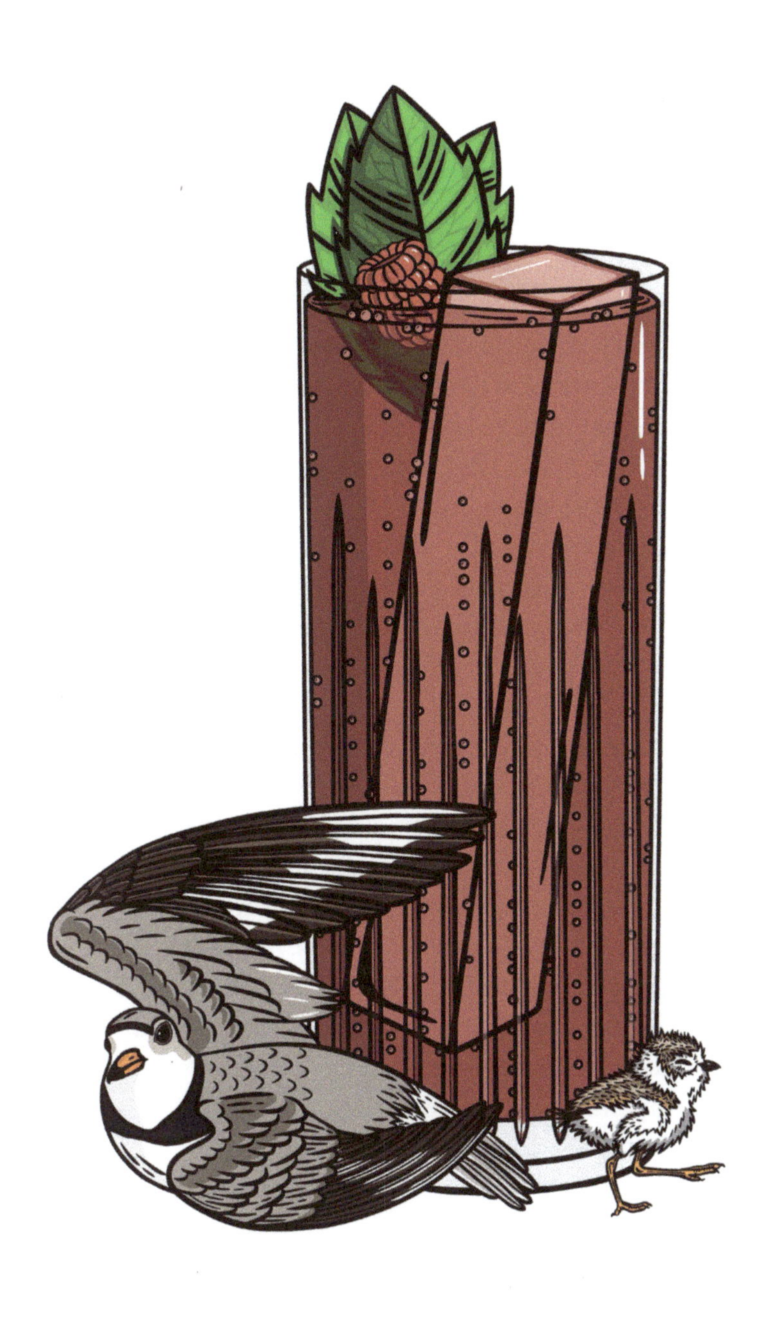

Red Herring Gull

How many species of birds are there in the world? For those of us who like keeping lists, it's an important question. The answer is 11,145. Or 11,250. Or maybe 11,195. It all depends on who you ask.

There are three main bird species lists: the Clements Checklist (maintained by the Cornell Lab of Ornithology and used on eBird), the International Ornithological Congress (IOC) World Bird List, and the BirdLife International Checklist. Individual regions also have their own lists, such as the checklists of the American Ornithologists' Union and British Ornithologists' Union. The differences between all these lists are relatively small, but they can become contentious.

Take the herring gull, for example (formerly *Larus argentatus*). Once considered a single, widespread species, it has now been split into four by all three major checklists: the American herring gull, the European herring gull, the Vega gull, and the Mongolian gull. This concordance only just happened in 2024. Before that, the Clements, IOC, and BirdLife lists identified one, three, and four species, respectively. Each list is compiled by a committee of respected ornithologists who are all looking at the same data. How do they end up so different?

To answer this, we must first ask a much more difficult question: what *is* a species? Bring this up over beers with a group of biologists and things might get rowdy. *Generally*, we define separate species as two groups that are too different to reproduce. But how do we know whether or not they can reproduce if they live in different places? We can use DNA and make a judgment call based on how different they are genetically, but how different is different enough? What if they have almost identical DNA, but still don't interbreed (see the Doctor Bird, pg. 81)? It's easy to see how the experts might end up disagreeing about 100 species out of 11,000.

Fortunately, there is currently an effort to reconcile all these lists into one. The International Ornithologists' Union is working with all of these organizations to put together a Working Group Avian Checklist that will—in theory—be *the* definitive list. At the time of writing, it's due out any day. I suspect there will still be four species of herring gull...but I might wait and see before I update my life list.

This drink looks and tastes like a spicy grapefruit shandy, but the hoppy flavor is just a red herring! Nonalcoholic beer can taste remarkably close to the real thing—I recommend the offerings from Athletic Brewing Company.

Red Herring Gull

4 oz nonalcoholic IPA
4 oz grapefruit soda
¼ oz hot honey syrup
1 sprig rosemary

Add honey syrup to the bottom of a chilled highball glass. Top with cold nonalcoholic IPA and grapefruit soda. Stir briefly. Garnish with a sprig of fresh rosemary.

HOT HONEY SYRUP

½ cup Mike's Hot Honey or a similar product
½ cup hot water

Combine honey and water and stir until dissolved. Let cool before using. Store in the refrigerator. Try it in a spicy Bee's Knees.

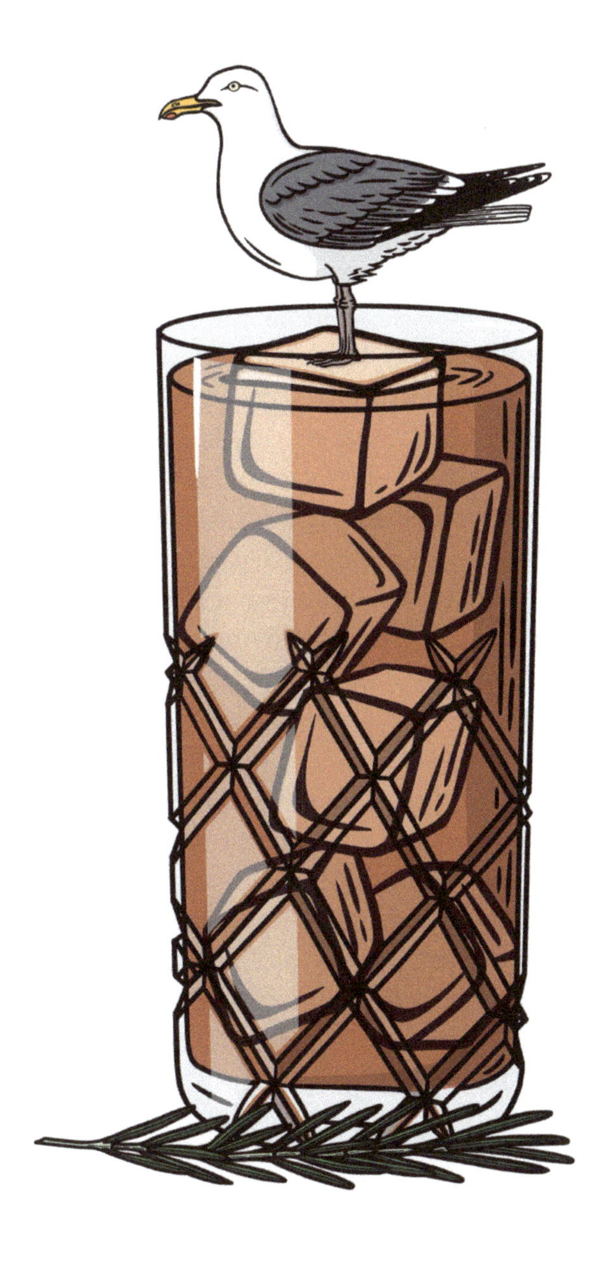

Tug-of-Warbler

Bird migration is a fascinating subject. Migratory birds seem to be born knowing where to go and when to leave. They travel thousands of miles on instinct alone. Meanwhile, I use Google Maps to get to the grocery store. Can directions really be in your DNA?

A lot of what we know about this subject comes from the Eurasian blackcap (*Sylvia atricapilla*), a warbler that breeds throughout Europe and western Asia and migrates to Africa and the Mediterranean coast for the winter. There is a line between 10 degrees and 11 degrees latitude, running right through Norway, Denmark, Germany, Switzerland, and northern Italy, that cuts the blackcap's European range neatly in half. Blackcaps that breed west of this line migrate to west Africa and the Iberian peninsula, whereas those that breed east of the line migrate to east Africa and the eastern part of the Mediterranean. Scientists took birds from one side of the line to the other and found that they still try to migrate in their original direction. And if two birds from opposite sides of the line have chicks, those offspring will try to migrate in an *intermediate* direction, halfway between the two.[37] So while environmental factors like temperature and day length may tell Blackcaps when to migrate, the direction they go does appear to be encoded in their genes.

I'm sure nocturnal migrants like the Eurasian blackcap would knock back a few espresso shots before they left if they could. Espresso and tonic may sound a bit strange, but don't knock it until you try it—especially combined with lavender and vanilla.

Tug-of-Warbler

4 oz tonic water
1 ½ oz espresso, cooled
¼ oz lavender syrup
1 dash vanilla extract

Combine espresso, lavender syrup, and vanilla in the bottom of a rocks glass and stir. Fill the glass with ice and top with tonic water. Garnish with a sprig of fresh lavender.

Whydah Hell Not?

Birds love big, flashy tails. From the massive plumes of the Indian Peafowl to the bizarre, curving racket feathers of the marvelous spatuletail hummingbird, fancy tails are clearly all the rage among male birds. Why do the ladies love it so much?

Take the male long-tailed paradise whydah (*Vidua paradisaea)* of sub-Saharan Africa. During the mating season, this tiny finch sports a glossy black tail that can be several times his own body length. Growing those massive feathers every year takes a lot of energy, as does keeping them in good condition. And it's much, much easier to be caught and eaten with a long, cumbersome tail. The whydah is showing potential mates that he is a strong, high-quality male who is able to spend resources on growing, maintaining, and surviving with that massive tail.

Similarly, cocktail drinkers love big, flashy garnishes. One of my favorite bird/cocktail crossovers is an essay by famous bartender Sasha Petraske called "Consider the Peacock."[38] Elaborate cocktail garnishes, Petraske says, are like long tails in birds: indicators of quality. If a bartender takes time to craft a beautiful garnish, they probably took time and care with the drink. If the herbs and fruit in the garnish are fresh, then the ingredients in the drink must be as well.

Make Myna Double

With that in mind, make sure you use only the freshest basil to garnish this tropical mocktail, and/or get flashy and use pineapple fronds, a lime wheel, or a little paper umbrella. Whydah hell not?

Whydah Hell Not?

2 oz pineapple juice
¾ oz lime juice
¾ oz agave nectar
3 oz club soda
6 basil leaves
Tajin, for rim

Run a slice of lime along the rim of a highball glass and dip it in Tajin. Fill the glass with ice. Add basil and agave nectar to the bottom of a shaker and muddle gently. Add pineapple juice and lime juice. Fill the shaker with ice and shake until chilled. Strain into the prepared glass and top with club soda. Garnish with fresh basil.

References

Gin

1. Velando, A., Beamonte-Barrientos, R., & Torres, R. (2006). Pigment-based skin colour in the blue-footed booby: An honest signal of current condition used by females to adjust reproductive investment. *Oecologia, 149*(3), 535–542.

2. Spottiswoode, C. N., & Wood, B. M. (2023). Culturally determined interspecies communication between humans and honeyguides. *Science, 382*(6675), 1155–1158.

3. Spottiswoode, C. N., Stryjewski, K. F., Quader, S., Colebrook-Robjent, J. F. R., & Sorenson, M. D. (2011). Ancient host specificity within a single species of brood parasitic bird. *Proceedings of the National Academy of Sciences, 108*(43), 17738–17742.

Rum

4. Theron, E., Hawkins, K., Bermingham, E., Ricklefs, R. E., & Mundy, N. I. (2001). The molecular basis of an avian plumage polymorphism in the wild: A melanocortin-1 receptor point mutation is perfectly associated with the melanic plumage morph of the bananaquit, Coereba flaveola. *Current Biology, 11*(8), 550–557.

5. Mundy, N. I., Badcock, N. S., Hart, T., Scribner, K., Janssen, K., & Nadeau, N. J. (2004). Conserved genetic basis of a quantitative plumage trait involved in mate choice. *Science, 303*(5665), 1870–1873.

6. U.S. Geological Survey. (2022, May 25). *A climate change canary in the coal mine – The endangered Hawaiian honeycreepers*. U.S. Department of the Interior, U.S. Geological Survey. www.usgs.gov/news/featured-story/a-climate-change-canary-coal-mine-endangered-hawaiian-honeycreepers

7. Judy, C. D., Graves, G. R., McCormack, J., Stryjewski, K. F., & Brumfield, R. T. (2025). Speciation with gene flow in an island endemic hummingbird. *PNAS Nexus, 4*(4), pgaf095.

8. Jordan, E. A., & Areta, J. I. (2020). Biosonic mechanical wing songs and complex kinematics in aerial displays of the subtropical doradito (Pseudocolopteryx acutipennis). *Integrative and Comparative Biology, 60*(5), 1173–1187.

9. Bostwick, K. S., & Zyskowski, K. (2001). Mechanical sounds and sexual dimorphism in the crested doradito. *The Condor: Ornithological Applications, 103*(4), 861–865.

10. Latta, S. C., & Wunderle, J. M., Jr. (1996). Ecological relationships of two Todies in Hispaniola: Effects of habitat and flocking. *The Condor: Ornithological Applications, 98*(4), 769–779.

11. Brightsmith, D. J., & Muñoz-Najar, R. A. (2004). Avian geophagy and soil characteristics in southeastern Peru. *Biotropica, 36*(4), 534–543.

12. Powell, L. L., Powell, T. U., Powell, G. V. N., & Brightsmith, D. J. (2009). Parrots take it with a grain of salt: Available sodium content may drive collpa (clay lick) selection in southeastern Peru. *Biotropica, 41*(3), 279–282.

13. Brightsmith, D. J., Hobson, E. A., & Martinez, G. (2017). Food availability and breeding season as predictors of geophagy in Amazonian parrots. *Ibis, 160*(1), 112–129.

14. Carroll, N. R., Chiappe, L. M., & Bottjer, D. J. (2019). Mid-Cretaceous amber inclusions reveal morphogenesis of extinct rachis-dominated feathers. *Scientific Reports, 9,* 18108.

15. Bleiweiss, R. (1987). Development and evolution of avian racket plumes: Fine structure and serial homology of the wire. *Journal of Morphology, 194*(1), 23–39.

16. Murphy, T. G. (2007). Lack of melanized keratin and barbs that fall off: How the racketed tail of the turquoise-browed motmot Eumota superciliosa is formed. *Journal of Avian Biology, 38*(2), 139–143.

17. Tattersall, G. J., Andrade, D. V., & Abe, A. S. (2009). Heat exchange from the toucan bill reveals a controllable vascular thermal radiator. *Science, 325*(5939), 468–470.

Tequila

18. Federal Aviation Administration. (2025, January 3). *Frequently asked questions and answers.* U.S. Department of Transportation, Federal Aviation Administration. www.faa.gov/airports/airport_safety/wildlife/faq

Vodka

19. Samuel, M. D., Liao, W., Atkinson, C. T., & LaPointe, D. A. (2020). Facilitated adaptation for conservation – Can gene editing save Hawaii's endangered birds from climate driven avian malaria? *Biological Conservation, 241,* 108390.

20. Klicka, J., Burns, K., & Spellman, G. M. (2007). Defining a monophyletic Cardinalini: A molecular perspective. *Molecular Phylogenetics and Evolution, 45*(3), 1014–1032.

21. Birkhead, T. R., Thompson, J. E., & Montgomerie, R. (2018). The pyriform egg of the Common Murre (Uria aalge) is more stable on sloping surfaces. *The Auk: Ornithological Advances, 135*(4), 1020–1032.

22. Birkhead, T. R., Thompson, J. E., Cox, A. R., & Montgomerie, R. D. (2021). Exceptional variation in the appearance of Common Murre eggs reveals their potential as identity signals. *The Auk: Ornithological Advances, 138*(4), ukab049.

23. Fleming, F. (Ed.). (2016). *The man with the golden typewriter: Ian Fleming's James Bond letters.* Bloomsbury Publishing.

Whiskey

24. Rutz, C., Sugasawa, S., van der Wal, J. E. M., Klump, B. C., & St Clair, J. J. H. (2016). Tool bending in New Caledonian crows. *Royal Society Open Science, 3*(8), 160439.

25. Taylor, A. H., Elliffe, D., Hunt, G. R., & Gray, R. D. (2010). Complex cognition and behavioural innovation in New Caledonian crows. *Proceedings of the Royal Society B: Biological Sciences, 277*(1694), 2637–2643.

26. Maderspacher, F. (2022). Flightless birds. *Current Biology, 32*(20), R1155–R1162.

27. Hume, J. P., & Martill, D. (2019). Repeated evolution of flightlessness in Dryolimnas rails (Aves: Rallidae) after extinction and recolonization on Aldabra. *Zoological Journal of the Linnean Society, 186*(3), 666–672.

28. Katz, B. (2019, May 13). How evolution brought a flightless bird back from extinction. *Smithsonian Magazine*. www.smithsonianmag.com/smart-news/how-evolution-brought-flightless-bird-back-extinction-180972166

Other Base Spirits

29. Shakya, S. B., Lim, H. C., Moyle, R. G., Rahman, M. A., Lakim, M., & Sheldon, F. H. (2019). A cryptic new species of bulbul from Borneo. *Bulletin of the British Ornithologists' Club, 139*(1), 46–55.

30. Harshman, J., Braun, E. L., Braun, M. J., & Yuri, T. (2008). Phylogenomic evidence for multiple losses of flight in ratite birds. *Proceedings of the National Academy of Sciences, 105*(36), 13462–13467.

31. Davies, N. B. (1983). Polyandry, cloaca-pecking and sperm competition in dunnocks. *Nature, 302*(5906), 334–336.

Mocktails

32. Prakash, V., Green, R. E., Pain, D. J., Ranade, S. P., Saravanan, S., Prakash, N., Venkitachalam, R., Cuthbert, R., Rahmani, A. R., & Cunningham, A. A. (2007). Recent changes in populations of resident Gyps vultures in India. *Journal of the Bombay Natural History Society, 104*(2), 127–133.

33. Frank, E., & Sudarshan, A. (2024). The social costs of keystone species collapse: Evidence from the decline of vultures in India. *American Economic Review, 114*(10), 3007–3040.

34. Galligan, T. H., Green, R. E., Wolter, K., Taggart, M. A., Duncan, N., Mallord, J. W., Alderson, D., Li, Y., & Naidoo, V. (2022). The nonsteroidal anti-inflammatory drug nimesulide kills Gyps vultures at concentrations found in the muscle of treated cattle. *Science of the Total Environment, 807*(2), 150788.

35. McCracken, K. G., Wilson, R. E., McCracken, P. J., & Johnson, K. P. (2001). Are ducks impressed by drakes' display? *Nature, 413*(6852), 128.

36. de Framond, L., Brumm, H., Thompson, W. I., Drabing, S. M., & Francis, C. D. (2022). The broken-wing display across birds and the conditions for its evolution. *Proceedings of the Royal Society B: Biological Sciences, 289*(1971), 20220058.

37. Helbig, A. J. (1991). Inheritance of migratory direction in a bird species: A cross-breeding experiment with SE- and SW-migrating blackcaps (Sylvia atricapilla). *Behavioral Ecology and Sociobiology, 28*, 9–12.

38. Petraske, S., & Moger-Petraske, G. (2016). *Regarding cocktails*. Phaidon Press.

Cocktail Life List

Check off what you've made! Recipes are in taxonomic order by bird species (see Grain of Saltator, pg. 140).

- ❑ Get Rheal
- ❑ Daiqkiwi
- ❑ Apple Eider
- ❑ Lame Duck
- ❑ Boom Chachalaca
- ❑ Bobwhite Russian
- ❑ Sage-Grouse
- ❑ Ptarmigimlet
- ❑ Bamboo Partridge
- ❑ Sazeracket-tail
- ❑ Doctor Bird
- ❑ Suffering Bustard
- ❑ Piña Koel-ada
- ❑ Paloma
- ❑ Dove Potion No. 9
- ❑ Rusty Rail
- ❑ Plover Club
- ❑ Blood and Sandpiper
- ❑ Red Herring Gull
- ❑ Jaeger Bomb
- ❑ Moscow Murre
- ❑ Bella Loona
- ❑ Fulmargarita
- ❑ Stork Club
- ❑ Booby Trap
- ❑ Scarlet Rye-bis
- ❑ Bitter-n
- ❑ Hoatzinfandel
- ❑ Culture Vulture
- ❑ Kite-pirinha
- ❑ Free-for-Owl
- ❑ Hoopoe Coupe-o
- ❑ Maharaja's Kookaburra-Peg
- ❑ Hot Tody
- ❑ Bee-Eater's Knees
- ❑ Jacamartinez
- ❑ Toucan Play at That Game
- ❑ Honeyguide
- ❑ Jungle Bird

- ❑ Macawsmopolitan
- ❑ Doradito Mojito
- ❑ Mai Tyrant
- ❑ Negrito Negroni
- ❑ Rob Royal Flycatcher
- ❑ Compulsive Lyrebird
- ❑ Cardinale
- ❑ Sidecardinal
- ❑ Vangastura Sour
- ❑ Pitohui Punch
- ❑ Philadelphia Fish Crow Punch
- ❑ Gin & Tit
- ❑ Queen's Lark Swizzle
- ❑ Bulbullini
- ❑ Purple Martini
- ❑ Tug-of-Warbler
- ❑ Coconuthatch
- ❑ Porn Starling Martini
- ❑ Tequila Sunbird
- ❑ Whydah Hell Not?
- ❑ LBB
- ❑ Brambling
- ❑ Blue Hawaiian Honeycreeper
- ❑ Appletini'iwi
- ❑ Vesper Sparrow
- ❑ Mad Cowbird
- ❑ Grain of Saltator
- ❑ Bananaquit
- ❑ Yellow Bird

Acknowledgments

Thank you to the Boston hospitality scene for being so warm and welcoming. Countless bartenders and Instagrammers have provided inspiration to me over the years, all of which has been poured into this book. The Miller High Life sidecar for a Porn Star Martini riff apparently buried itself in my brain after a visit to The Wig Shop in Boston—credit for the idea goes to them.

Particular thanks to Brian Hoefling and Alison Faust for their detailed feedback on the manuscript, to Joey Faust for suggesting the cocktail life list, and to Adam Cohen (a.k.a. @mrmuddle) for testing out some recipes.

I want to thank my parents for always encouraging my interests, even when they transferred from something already impractical and not particularly lucrative (ornithology) to something even more impractical and less lucrative (cocktail Instagramming). They threw themselves into both pursuits with enthusiasm, adopting my hobbies as their own and giving me two more people to discuss them with. Mom, I know you would have adored every recipe, fact, and terrible pun in this book, and I wish so much that you were here to read it.

Thank you to my son Luke for his boundless enthusiasm for this project, despite how often I had to work on it instead of playing with him, and to my dog Moresby for his (literally) constant companionship.

The biggest thank you goes to my husband Tommy, for supporting me and encouraging me, making me braver, and taking care of so many things when my terrible time management skills get the better of me. I'd be a mess without you. I love you!

And finally, I'd like to thank my academic advisors, professors, and colleagues for the guidance and knowledge they imparted...but mostly for occasionally driving me to drink.

About the Author

Katie Faust Stryjewski is a cocktail photographer, recipe creator, Instagram influencer, former ornithologist, and (as of the publication of this book) professional illustrator. She fell in love with birds while growing up in southern Louisiana. She attended Louisiana State University, where she worked at the Museum of Natural Science on a project examining the genetics of a hybrid zone between two species of Jamaican streamertail hummingbirds (Doctor Bird, pg. 81). She then moved to Boston for graduate school at Boston University, where she studied the genomics of rapid speciation in a group of estrildid finches (genus *Lonchura*) from Papua New Guinea and Australia, as well as the genetics of egg shape in honeyguides (Honeyguide, pg. 60). She did a post-doc at Harvard examining variation in avian retinal morphology (Hawkward Position, pg. 222). She then completely switched gears and became a cocktail content creator, publishing her work on Instagram (@ garnish_girl) and her blog Garnish (www.garnishblog.com). Her first book, *Cocktails, Mocktails and Garnishes from the Garden*, was released by Mango in 2021. She currently lives in Boston with her husband, son, and dog.

Index

A

B

C

D

E

F

G

H

I

J

K

L

M

N

R

S

T

U

V

W

Y

Mango Publishing, established in 2014, publishes an eclectic list of books by diverse authors—both new and established voices—on topics ranging from business, personal growth, women's empowerment, LGBTQ studies, health, and spirituality to history, popular culture, time management, decluttering, lifestyle, mental wellness, aging, and sustainable living. We were named 2019 *and* 2020's #1 fastest growing independent publisher by *Publishers Weekly*. Our success is driven by our main goal, which is to publish high-quality books that will entertain readers as well as make a positive difference in their lives.

Our readers are our most important resource; we value your input, suggestions, and ideas. We'd love to hear from you—after all, we are publishing books for you!

Please stay in touch with us and follow us at:

Facebook: Mango Publishing
Twitter: @MangoPublishing
Instagram: @MangoPublishing
LinkedIn: Mango Publishing
Pinterest: Mango Publishing
Newsletter: mangopublishinggroup.com/newsletter

Join us on Mango's journey to reinvent publishing, one book at a time.

www.ingramcontent.com/pod-product-compliance
Lightning Source LLC
Jackson TN
JSHW040832090925
90166JS00001B/1

* 9 7 8 1 6 8 4 8 1 7 9 7 9 *